RIVER DAM CONSTRUCTION AND RESETTLEMENT OF AFFECTED VILLAGES

RIVER DAM CONSTRUCTION AND RESETTLEMENT OF AFFECTED VILLAGES

A Case Study of Two Villages in the Upper Krishna Valley: Vyahali and Chandavadi (District Satara)

PRAKASH R. SAWANT

INTER-INDIA PUBLICATIONS
D-17 RAJA GARDEN EXTN. NEW DELHI-110015 (INDIA)

Cataloging in Publication Data—DK

Sawant, Prakash R. (Prakash Ramchandra), 1942–

River dam construction and resettlement of affected villages.

Includes index.

1. Dhom Dam, India. 2. Dams—India—Maharashtra—Design and construction. 3. Maharashtra (India)—Population, Rural—Rehabilitation. 4. Villages—India—Maharashtra. 5. Kistna Valley, India. 6. Rivers—India—Maharashtra. 7. Maharashtra (India)—Rural development. I. Title.

First Published in India in 1985

Prakash Ramchandra **Sawant** (*b. 1942-*)

Published in India by:
M.C. Mittal
INTER-INDIA PUBLICATIONS
D-17, RAJA GARDEN EXTN.,
NEW DELHI-110015 (INDIA)
PHONE : 5413145

Printed in India by
Alka Enterprises
at Sunil Printers, Naraina, New Delhi-110028

Planning is not merely a question of drawing up a list of projects and finding resources to execute them. These two tasks, no doubt, form the base of planning.

Among the resources, thought is mostly given to the financial aspects and in fact, there is the danger of this aspect, rather overshadowing and suppressing the other aspects. The financial aspect is important but is far less important than the human aspect.

—JAWAHARLAL NEHRU

"The project on which the present report is based was funded by the Indian Council of Social Science Research. However, the responsibility for the facts stated, opinions expressed and conclusions reached is entirely that of the project director/author and not of the Indian Council of Social Science Research, New Delhi."

Comments on the Project Report from a Consultant of ICSSR

ON GOING through this research report, I find that the theme selected is such that it deserves to be studied in other similar localities where dam construction of some magnitude is under way. The technological aspects of construction and its intended benefits are known well enough and these are invariably highlighted by Government agencies, political leaders and the press. But little is known about what a dam construction means to the community that bears the impact, which section of it benefits and which one suffers. While this modern technology aims to promote economic development, it also implies environmental change and the necessary shift of population from catchment area to elsewhere. A mute section of humanity is uprooted. It is this aspect of man's response to a man-made environmental and locational change that merits the attention of those who research in "Human Geography". It is from this point of view that the present study is welcome. Mr. Sawant who belongs to this area known his home tract well and has treated his theme with concern, sympathy and understanding.

In his Chapter 2 he has stated his hypothesis well, given the aspects of rehabilitation in its magnitude, and Government policy and implementation, and the useful but little known distinction between fully and partially affected population and their old and new homes, and the largely indifferent beneficiaries in the command area. All this is done in the background of the physical setting of the Upper Krishna Valley. The data painstakingly generated is good and competent. The sample studies of the two villages in their old and new sites are good. Mr. Sawant gives a balance attention to the land and the people. The villagers' response to their old and new environment is well brought out. The "shift" (Chapter 8) of the population as perceived by Government personnel, public leaders and the affected community is well studied though one wishes that he had studied the affected community's response in greater depth. He has also tried to evaluate the gains accruing to the beneficiary community living in the command area. Chapter 10 and 11 aim at identifying guidelines for rehabilitation of displaced families and how Government agencies could be more competent and sympathetic. He rightly ends the study with the suggestion that similar studies for the dam construction affected villages and their inhabitants need to be carried out.

In handling his theme, Mr. Sawant shows what is happening today in our country to young and keen researchers who have to live in academic solitude, with no exposure to research experience, much less to the right use of research methodology and techniques. In spite of these handicaps, Mr. Sawant has put in his best to make his study meaningful. By sophisticated standards, his study may be almost pedestrian. It would have improved with rigorous application of statistical and quantitative techniques. For instance, he has lost a good opportunity in comparing the characteristics of old villages and their new counterparts with the aid of selected indicators. Again he could have possibly answered that important question : Whether the village community continues to be caste ridden in their localities in new sites, although he does bring out that the Harijan castes reside still outside the new village.

But Prof. Sawant's effort has its merits. It is free from that jargon and blind application of Western concepts which I see in many of our Ph.D. theses and MSS submitted to ICSSR. Further he has a freshness of approach and he has tried to see reality as it exists and operates. This is what, in my opinion, makes his study meaningful and useful. I appreciate his efforts, and altogether reading his report has gained me a better insight in the human problems in rehabilitation.

27th December, 1984

Preface

THE ECONOMIC planning in India, has been geared up since 1952, and the process is vigorously and rightly focused on the agricultural development. In an effort of transformation of rural landscape through irrigation and hydroelectricity projects, some villages are entirely shifted off their base, losing their identity and the cultural environment possessed by them. The process of rehabilitation of these villages at present includes giving the villagers alternative land and/or monetary compensation. This economic planning irrespective of a thought to securing the cultural entity and social cohesiveness of the settlements, may not prove to be conductive for environmental conservation in that particular landscape.

These problems have been surfacing frequently wherever river/dams are being proposed and constructed at different locations all over India. The problem of shifting and rehabilitation of settlements is required to be given a thorough consideration. River valley development projects are not only formed by construction of irrigation projects. They should be formed by social, cultural and economic environmental considerations. Since the river valley projects are indispensable components of the regional planning exercises in India, identification of the rehabilitation problems and measures towards their amelioration have become significant.

The immediate aim of the study is to assess the total affect of an irrigation project on rehabilitation. The study relates to the recently constructed Dhom Dam across the Krishna River in its upper reaches. In this study, an attempt is made to present the aspects of changes in landuse, social structure, economic patterns and the population mobility with reference to selected affected villages by Dhom Project. It is my hope that this report will be considered as a fact finding attempt and will be used for balanced development of "River Valley Projects and Rehabilitation Planning". The investigation work of the people affected by Krishna Dhom Dam was started in January 1982, and finished in January 1984.

I would like to express my gratitude to the Indian Council of Social Science Research, New Delhi-110001, for awarding the Council's grant of Rs. 8,767 to work on a theme of my choice.

I take this opportunity to record my grateful thanks to Dr. C.D. Deshpande, Pune and Dr. M.D. Tawde, Shivaji University, Kolhapur, for their encouragement and moral support, valuable guidance and suggestions for completion of this work.

I owe a debt of gratitude to Mr. Prataprao Bhosale, (Former Minister for Rural Development and Rehabilitation), Minister of Rural Development, Govt. of Maharashtra; Dr. Baba Adhav, Secretary, Maharashtra State Irrigation Project affected "Shetakari Parishad, Pune; Shri D.S. Telsinge, Asstt. Executive Engineer, Khanher Land Acquisition and Rehabilitation Office, Satara; Dr. A.A. Kadam, Principal, Kisan Veer Mahavidyalaya, Wai; and members of the staff of my college, for their valuable suggestions in this work.

I would like to thank, Prof. V.S. Choudhari, and V.L. Jadhav (investigators), for their assistance in collecting and processing the data and to Mr. B.B. Dhamal (Mandal Nirikshak) L.A. and Irrigation Office Wai, and Shri. A.V. Deshpande, District Rehabilitation Office, Satara, for statistical help and advice given to me in relation to Krishna Dhom Rehabilitation work. My thanks are also due to various officials connected with the Krishna Dhom Dam Project, other government departments, and Sarapanch of the Grampanchayat and all the village people of Dhom, Vyahali, (Kadegaon—Group of Grampanchayat), Chandavadi (Group Grampanchayat of Khadki), with whom I had long discussions during the course of my field work), for assistance, and co-operation given to me for collecting the data.

I also thank to the typists, Shri Arvind Bhoot (Wai), and Shri S.A. Pawar (Anewadi, Tal. Wai).

I am thankful to my wife Pavitra, who besides her domestic chores helped me at every stage of the work.

PROF. PRAKASH R. SAWANT,
(Project Director)
Department of Geography,
Kisan Veer Mahavidyalaya,
Wai (Dist. Satara)

Contents

List of Figures

List of Tables

List of Illustrations

OLD VYAHALI

20. A house
21-22. Unaffected settlement of upward part of old Vyahali
23. Rice : was a prime crop : Rice threshing
24. A part of settlement

OLD ENVIRONMENT

25. The natural beauty which forever lost under the "Dhom lake"

OLD CHANDAVADI

26. Settlement location
27. Settlement location
28. Walki and Krishna "Sangam" in Dhom Lake
29. Location of old Velang "Cattle Market Centre" lost under the sand and mud of the dam

NEW VYAHALI
(Within the command area village : Kadegaon, Tal. Wai)

30. The lower part of "Vairat Gad" range : now acquired and covered for settlement of New Vyahali
31. New Settlement from Wai-Satara road
32. "Vairat Gad": (Fort of Virat Raja") and the new Vyahali settlement
33. "Bhairav Nath Temple
34. Eastern part of the Settlement
35. Well : A source of drinking water
36. Central Part of the Settlement
37. Unsettled part of the village area near to Wai-Satara road
38. Eastern part of the Settlement and agricultural farm
39. A Lane of Western Section
40. Right Bank Canal at the South of the new settlement
41. B.C. Settlement
42. Source of Canal water : Closed agricultural land to new settlement part

JAMB MAL FOR "NEW CHANDAVADI"

43. Bhuinj Sugar Factory
44. Mahadeo Temple and the barren land of "Jamb Mal"

SOME LANDSCAPES OF UPPER KRISHNA VALLEY

List of Appendices

1

Introduction

The Problem

WITH the dawn of the planning era, the nation has had several achievements in its economic progress. One of the most striking achievements is the development of irrigation and power through many dam constructions, large and small.

The construction of a dam implies loss of terrain in its catchment area, and gain in agricultural productivity in the command area. In both these areas, there are geographical aspects which merit a close attention, especially from the point of view of human welfare at the micro-level. The present study aims at identifying the significance of the impact of a dam construction on the affected regional communities. The study relates to the recently constructed Dhom Dam across the Krishna river in its upper reaches. More specifically it relates to the people of 29 affected villages.

Hypotheses

For the present study which focuses its attention to change in geographical location and environment and the human response to the change. The following hypotheses were framed and tested :

(*i*) The shift to new sites is not easy.

(*ii*) The persons displaced are reluctant to leave their old homes and accept to new environmental settings even though they are offered by the government several facilities and incentives.

(*iii*) The shift to new village sites tends to introduce the persisting pattern of the village residential structure with its traditional division of caste-based wadas and gallies.

(*iv*) The landuse pattern in the new villages tends to be different from the one in the original villages under the influence of the new environmental and economic setting.

(*v*) The overall effect of the shift is change in the pattern and life style of the people which is different from that in the old villages.

(*vi*) Dam construction introduces changes in the ecological setting both in the catchment and command areas; these changes need to be taken into account for the purposes of planning and development in both the areas.

Methodology, Techniques and Data Sources

This is an empirical study of the Upper-Krishna valley and its settlements with special reference to the displaced villages, their new sites, and the affected communities. Suitable cartographic and statistical techniques have been used for the purpose of spatial analysis. The analysis includes both the more stable aspects of landuse in its natural environment and the more dynamic human activities in their social and economic settings.

Collection of Primary and Secondary Data

The work of field survey and collection of data was completed with the help of investigators under the guidance of the Project Director. Schedules were prepared and interviews based on the questionnaire were taken with the heads of the sample families. Copies of relevant questionnaires were also sent by post to the officials who were in charge of rehabilitation work. Some officials and local leaders were interviewed.

The secondary data was collected from the published reports and official records of the Revenue and Irrigation Departments.

Processing of Data and Sampling Techniques

From the various sources of data, the Project Director with the help of the investigators processed and analysed the data. The primary data is collected by selecting sample families of 20 per cent of Vyahali and 40 per cent of Chandavadi families. New Vyahali and New Chandavadi (Chapter 6 and 7), were the selected villages for study. Both the villages became affected villages by the Dhom Dam as they were located at the very site of the dam.

To understand economic conditions, the villages were studied on the basis of landholdings and the landholders. Medium small farmers are those who have below 2.02 ha. (5 acres) of landholdings.

Random sampling techniques were employed. The families selected for the study as a sample from each village were stratified according to the landholdings, as given below ;

Category A : Up to 3 acres (Up to 1.21 ha.) of landholdings
Category B : 3 to 5 acres (1.21 to 2.02 ha.) of landholdings
Category C : 5 to 10 acres (2.02 to 4.04 ha.) of landholdings
Category D : 10 to 15 acres (4.04 to 6.06 ha.) of landholdings

Landless people affected by the Dhom Project were included and studied in the first category of "A". Small landholders were included in "A" and "B" groups. Medium group of landholders were included in the "C" group and big landholders were included in "D" group. The selected categories were formed into 'small', 'medium' and 'big' landholders, in order to understand their economic position at the rehabilitation sites.

The selected families were studied by dividing the family members into two age groups. Those above the age group of 16 years who were eligible to work and earn for themselves and for their families and the other below the age of 16 years, mainly as are dependents.

No landholders holding more than 6.06 ha. (15 acres) of agricultural land were observed within the selected villages and even in all the villages affected by Dhom Project. Even families with 'D' category of holding 4.04 to 6.06 ha. (10 to 15 acres) of agricultural land were rare.

Secondary and primary data, so gathered, is processed cartographically and with the application of questionnaire method.

Results and their Significance

The results of this research study are calculated to be useful for the purposes of regional planning or replanning in the Upper Krishna Valley. It has also a methodological and practical usefulness, because the problems identified in this study and measures suggested towards their amelioration would be useful to those parts of India where such problems of rehabilitation are likely to come up in future. The results are indicated in the last topic of "General Observations and Conclusions".

2

Rehabilitation Dimensions, Government Policy and Community Response

The Magnitude of the Problem

THE magnitude of the rehabilitation problem can be understood against the background of development of irrigation projects in terms of number of projects, the extent of the added irrigated area, the number of villages and the people affected by dam construction.

In India, the following specific examples give an idea of displaced villages, with affected population due to irrigation project. Hirakud dam on Mahanadi (Orissa State) affected over 249 villages and 20,000 population. The Government of Orissa resettled these by providing maximum facilities only to 33 villages out of 249. In Madhya Pradesh, under the Rehand Project 105 villages were submerged. Under the Chambal Project (in Gandhi Sagar Dam) 228 villages were affected by losing their land under the dam. The "Srisailam Project", in Andhra Pradesh, which will be completed by September 1983, will submerge 100 villages displacing over one lakh people, 1,06,925 acres of land and 20,728 households, (Shatrugna, 1981).[1] In Maharashtra, for Koyna Project, a total of about 100 villages were affected involving a population of about 30,000 (Koyna dam affected 15 villages resettled in Satara district, 1 village in Ratnagiri district, 8 villages in Raigad district, 3 villages in Thana district and 2 villages in the Sangli district (Karve—Nimbkar 1969).[2] The Dudhganga Project (Kalammawadi Dam) Kolhapur district construction work was started in 1977-78 and it will be completed by 1987. Water will be stored in the lake by 1985-86. Under this project, there would be 9 fully affected villages. The benefited villages of Shirol and Hatkangale talukas have agreed to give their agricultural land for the 'Dudhganga' dam affected villages. Under the Warna Dam Project in Sangli district, there are 25 affected villages (16 fully and 9 partly affected villages) yet to be rehabilitated within the command area. The involved population under the Warna project is 4,730. (*Kesari*, Marathi Daily, 5th March, 1983).

Benefits

The economic benefits of added irrigation projects are too well known to stand a detailed discussion. Land productivity in the command area is raised. There are many studies of cost-benefit type by economists. There are some studies, like the one of Karve I., analysing the social aspects of the change. Now-a-days, attention is also being given to the problem of ecological change and conservation of natural resources. It is not known, however, if there are any studies even from the point of view of economies, as to understand who are the major beneficiaries of an irrigation project—the affected community or the unaffected and possibly better off sections of the society in the command area. It is most likely that the latter are the main beneficiaries. However, the purpose of our study is to direct our enquiry to the former—the affected community, its new environmental setting as against the old one, and how the change occasioned influences the community, what is their response and how well they are settled.

The Process of Dam Construction and Management of Catchment Area and Command Area

The general technique is well known. Hence a few aspects as a necessary preliminary background to our study need to be iterated. Such a project is entirely the responsibility of the State Government working through its Revenue and Irrigation Departments principally but the association of other connected departments and of the local leadership through the Zilla Parishad and Taluka Panchayat and Development Blocks is invariably maintained. The project and its time schedule is brought to the notice of the public through Government Gazette and mass media. The official agency to construct the dam as well as its still channels, canals and distributaries, tries to adhere to its time schedule but as is so well known succeeds seldomly because of several difficulties. The local public are warned of the progress, particularly relating to the impounding of waters and the consequent filling up of the expanse upstream of the dam. It is at this stage that the population of the fully and partly submerged villages are warned well before time and on the eve of the filling of the impounded waters to create a lake. Government agencies help the affected people with finance and more specifically for shifting their belongings, house structures, cattle, etc., to the newly allocated sites, by providing transport and advice. In the catchment area little attention has been paid so far to study the ecological disturbance caused by impounded waters, their effect on the soil, incidence of disease, etc., in the catchment area. Technically the command area is progressively developed through canals and distribution, according to the law of the land and the nature of soils. The chief aim is to add productivity to the land by irrigation waters. In practice, the layout of these water courses is subject to stresses and strains of local political leaders. It may be finally mentioned that the non-Government land project construction both in catchment and command areas is acquired by government under its rules, and compensation paid. From our point of view it is important to note that such a project means a spatial change in landscape and landuse. Catchment area water expanse and its new problems, and in the command area its new problems and in the command area water distribution and its impact, and above all the shift of population from the old sites to the new ones. Thus in the larger frame, both catchment and command areas present opportunities for their regional planning.

Rehabilitation : The Process

The villages which come under submergence of any irrigation project are affected villages. From such affected villages, the agricultural and settlement area is acquired by the government. All the affected villages are shifted and resettled in a new location of command area. For such affected villages, it becomes necessary again to acquire the agricultural land within the command area. Resettled villages are provided with all the necessary facilities within their new settlement locations. As per Land Acquisition Act, the Government is acquiring the land by fixing a ratio within the command and catchment area. The rules have been framed by giving compensation of land or alternative land for residential and for agricultural purposes to the affected landholders, considering the number of family members as a base for distribution of land. On the basis of total population of the new village, the amenities like school building, playground, roads, provision of drinking water, land for "Khalwadi" and many other such facilities are provided by the government.

For obtaining compensation and other personal benefits to the affected families and to the village, it is essential to contact government offices related to rehabilitation and irrigation work.

Finally, no irrigation project work is complete without judging it in relation to the opportunities of employment of capital and layout resources and rehabilitation work completed in the command area (Shatrugna, 1981).[3]

Government Policy Objectives

A note by the Chief Engineer, Koyna Project, expressed the responsibility of the government in the following words: ". . . the problem to be solved is primarily that of rehabilitation and not one of general betterment of the community as a whole. The general socio-economic betterment of people is a nation-wide problem and cannot be linked with the problem of rehabilitation. In rehabilitation, an attempt is made, as far as possible and practicable, to give the affected people some facilities and means of livelihood approximating to those they had in their original villages, but of course, with suitable improvement in hygienic and other living conditions in keeping with the present rate of advancement".

These are good thoughts, and one wishes their translation into action.

Government Policy and Implementation

In a reference to the problem of rehabilitation, the Chief Minister of Maharashtra stated on 24 September, 1982, that a committee for revising the Rehabilitation Act of 1976 would be appointed, and added that there are 1.19 lakh families affected out of which 47,000 families remain unrehabilitated (Ref. *Daily Aikya* : 24.9.1982). In general, it may be said that the earlier policy of paying compensation has been progressively improved upon by Government taking over the welfare aspects of rehabilitation. Thus, a Directorate of Rehabilitation and Settlement has been instituted to supervise and implement local plans for rehabilitation. There is co-ordination between government departments and local authorities like Zilla Parishads and local leaders. Of course there are administrative delays and more concernedly inadequate financial provision. But equally important is the response of the local community—both the affected and the non-affected sections.

Response of the Community

It is necessary to identify the response of both the affected and the non-affected sections because though the responses are different they need to be complementary and mutually supporting. The response of the affected community can be understood and sympathised. The change means, as stated earlier, a shift from an area and site to which they are traditionally and emotionally attached. Further, in settling down in a new place is to them a leap in the unknown ! Lack of education, poverty, inadequate skills and ignorance regarding what the shift implies, all these compound these difficulties. Hence their response to take the change is conditioned by indecision and anxiety. They literally drag their feet and are, therefore, out to shift to the new sites. It is important, therefore, that any such rehabilitation programme must be preceded by educating the community so that they may willingly accept the change and actively participate in it for their own benefit.

The response of the non-affected section is more relevant in the case of those who belong to the command area, because they receive the benefit of the project through increased productivity free from drought hazards. The interface in co-operation between the affected and the non-affected through government agencies, and secondly building up a new social homogeneity in the village which enfolds the new settlers. Experience shows that a non-affected person very unwillingly surrenders a part of his holding which is least fertile! Again, there are instances which new settlers are regarded with aversion. This is so even in the case of higher caste settlers, let alone the treatment of the Harijan settlers. This shows that it is also necessary to educate the non-affected person many of their better off due to irrigation availability towards cultivating the right social attitudes for absorbing the new comers in their environment.

Thus each project has its affected villages and affected population. And its non-affected economically better off villages and non-affected population which has a stable traditional background and prospects of seeing better times. The real prospect lies in mutual understanding and co-operation, aided by government policies and measures.

All this shows that we cannot gloss over rehabilitation work by only saying "त्यजेत् ग्रामं जनपदस्यार्थे···". The displaced villagers take to uncertain, tragic and unstable life, which is called a Marathi as "Upra" (उपरा). Dam affected villagers have to restart their life in the shadow of uncertainty and anxiety, while many, if not all, benefited villagers in the command area continue to live with stable, prosperous and productive irrigated agriculture. With a successful implementation and a plan for rehabilitation and with irrigation projects promoting irrigated farming on a large scale, the shortage of today, may turn tomorrow's surplus. As Pandit Jawaharlal Nehru said " . . . we want good men at the top to guide this great movement but we want even more good men at the village level. "Kurukshetra".[4] In our context more good men are necessary in administration, in political and social leadership and in the regional community of a project area.

REFERENCE

1. Shatrugna M. (1981), Unrehabilitated Poor of Sri Sailam Project, *Eco. & Pol Weekly*, XVI, 52, 26 Dec., 1981, Bombay.
2. Karve I., Nimbkar J. (1969), A survey of the people displaced through the Koyna Dam, (July 1965 to Jan. 1967), Deccan College, Pune.
3. Shatrugna M. (1981), Unrehabilitated Poor of Sri Sailam Project, *Eco. & Pol. Weekly*, XVI, 52, 26 Dec., 1981, Bombay.
4. Jawaharlal Nehru (1967), Kurukshetra, "Building up a new India, p. 6, A Symposium Publication Division, New Delhi.

3

The Physical and Environmental Setting of the Upper Krishna Valley and the Dhom Project

Physical Features

OUR STUDY area is a part of the Deccan plateau in Maharashtra. It is the small upper valley of a great river of sanctity—the Krishna. The valley is moulded into a typical basaltic topography of flat topped summits, step-like flanks and narrow valleys. The Krishna takes its source water at Kshetra Mahabaleshwar (1337 m. above sea level, 18° 1′ N and 73° 41′ E). Mahabaleshwar is about as the crow flies 64 kms., from the Arabian sea. From the source, the Krishna runs east for about 20 kms. in its course, until it reaches Dhom, where the Dhom dam is located. Then the river runs for 4 kms. eastwards, and reaches the town Wai. Wai is a taluka place and a market centre (701.0 m. A.S.L.) (see Fig. 3.0)

River Krishna enters into Wai taluka with its small tributary Walki Nadi (locally known as "Krishna Kamandalu Nadi") from the left in the upper part of the Dhom dam. This part is mountainous with steep slopes, cliffs and ridges. The general slope of the region is from NW to SE in Wai taluka. On the northern rim of the Krishna valley, there are the mountain ranges with the high peaks of Pandavgad (1273.85 m./4189′) which is maximum height in Wai tahsil (Fig. 3.1) and Mandhardeo (1252.80 m./4119′). The ranges extend from west to east up to Parkhandi and beyond these is another range "Dhamani Dongar" (1186.56 m./3900′). (Fig. 3.1). In the East, beyond village Kavathe, the hill of "Chavaneshwar Dongar" (1140.74 m./3755′) (Fig. 3.1) extends from north to south.

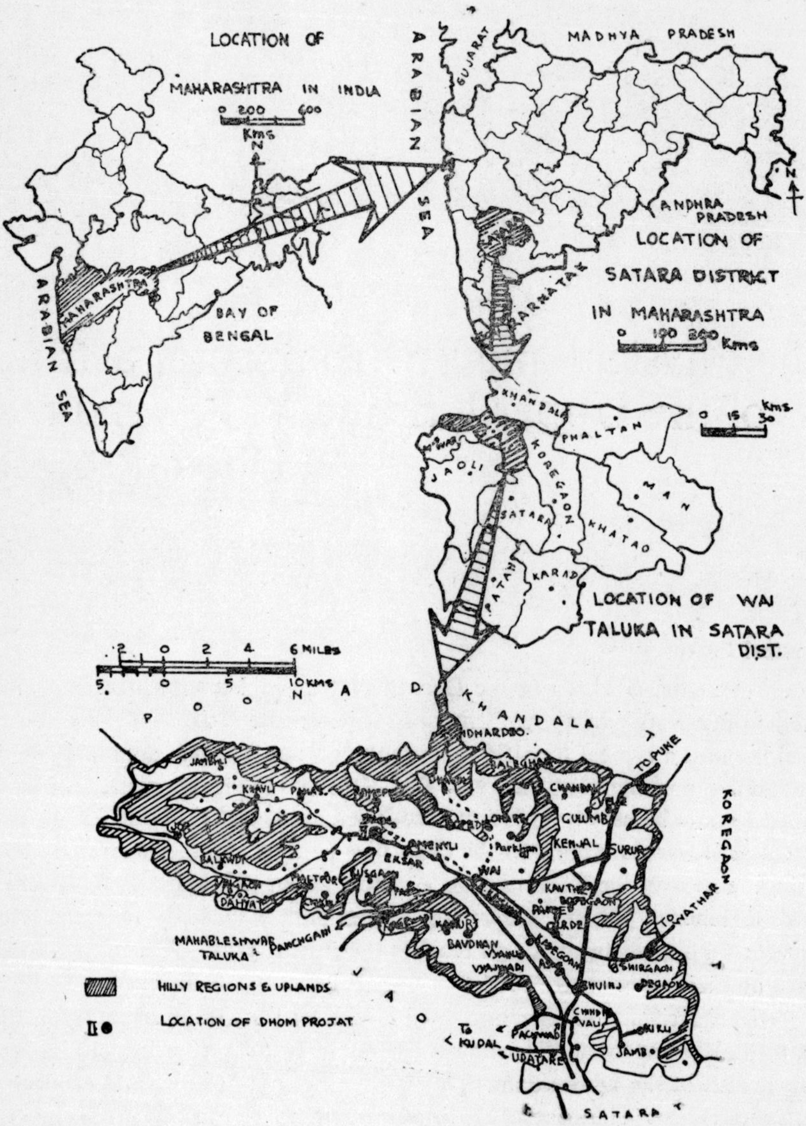

FIG. 3.0 : *Location of Wai Taluka in India and Location of Krishna Dhom Project*

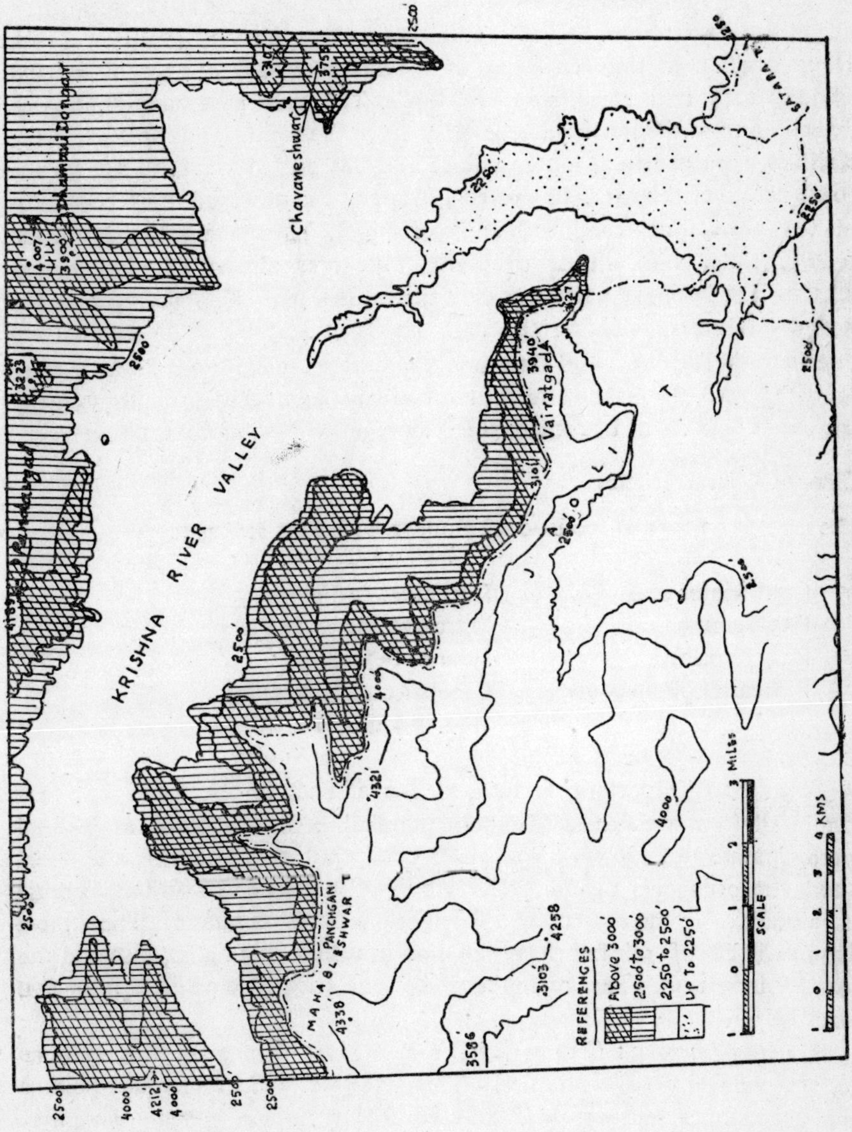

FIG. 3.1 : *Relief Map of Upper Krishna Valley in Wai Taluka*

On the south of the Krishna valley lie the Panchagani plateau and the Pasarni ghat range which merges the south boundary of Wai tahsil. It also extends from west to east. It has an ancient fort 'Vairatgad' with a height of 1198.68 m. A.S.L. (Fig. 3.1). The Krishna valley of the Dhom region is thus enclosed in a mountainous topography of flat summit tops and steep stepped slopes. The two mountain spurs, one extending from the north and the other from the south and approach the Krishna river course at Dhom village and this provides natural site for the construction of the dam. In this rugged area, it is observed that the fertility of the soil decreases with angle of slopes. This is on both sides of the river Krishna, with all its tributaries, becomes almost dry in summer season. The drainage pattern is dendritic in the area. Krishna receives the Kudali tributary on its right bank at about 3.2 kms. south from village Panchawad (Tal. Wai) and then enters in Satara tahsil and flows further south. (Fig. 3.2). In the stretch, there are no significant tributaries on the left bank, only small but seasonal streams, in the Wai area.

Climatic Characteristics

The area has a typical monsoonal climate with four seasons :

Cold winter season	December to February
Hot Summer	March to the end of May
Rainy season	June to September
Retreating monsoon	October and November
	(Fig. 3.4 & 3.5)

The most important feature of the climate is the rapid decrease in rainfall from west to east (Fig. 3.6). Mahabaleshwar records an annual average rainfall of 436 cm., Wai 71.00 cm., (710 mm.) and further east Bhuinj records only the annual variability also increases with the decrease in rainfall; so in the east of Wai drought hazard is common. Thus these characteristics of rainfall have an important bearing on landuse and the need for irrigation. The region can be climatically classified into two zones :

(*i*) *Western zone* : (From Wai-Dhom to Jor village at the extreme western village of Wai taluka). This is the wet and humid climatic area with steep slopes and narrow valleys, covered by dense forest. The area receives 145″ rainfall. Due to high rainfall of the zone, the location of Dhom project is suitable for collecting sufficient water for the lake.

(*ii*) *Eastern zone* of the valley in Wai region : (From Wai towards south-east up to Bhuinj and Panchawad village). This is a dry climatic zone

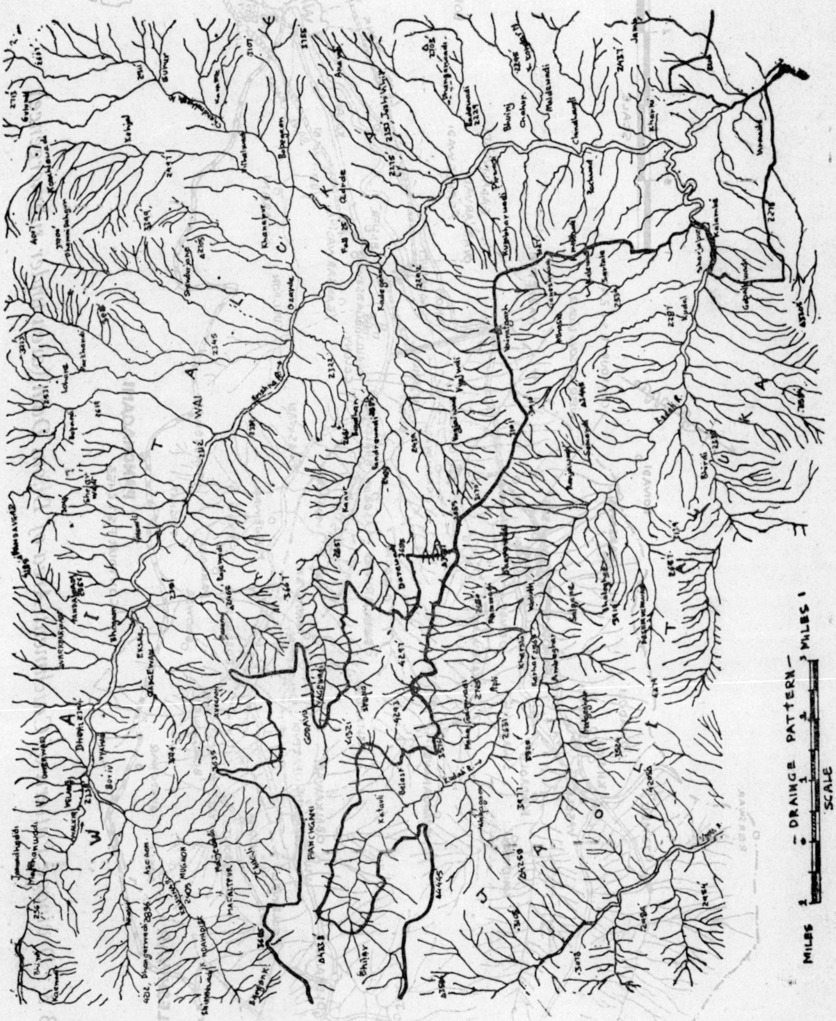

FIG. 3.2 : *Drainage Pattern of Upper Krishna Valley*

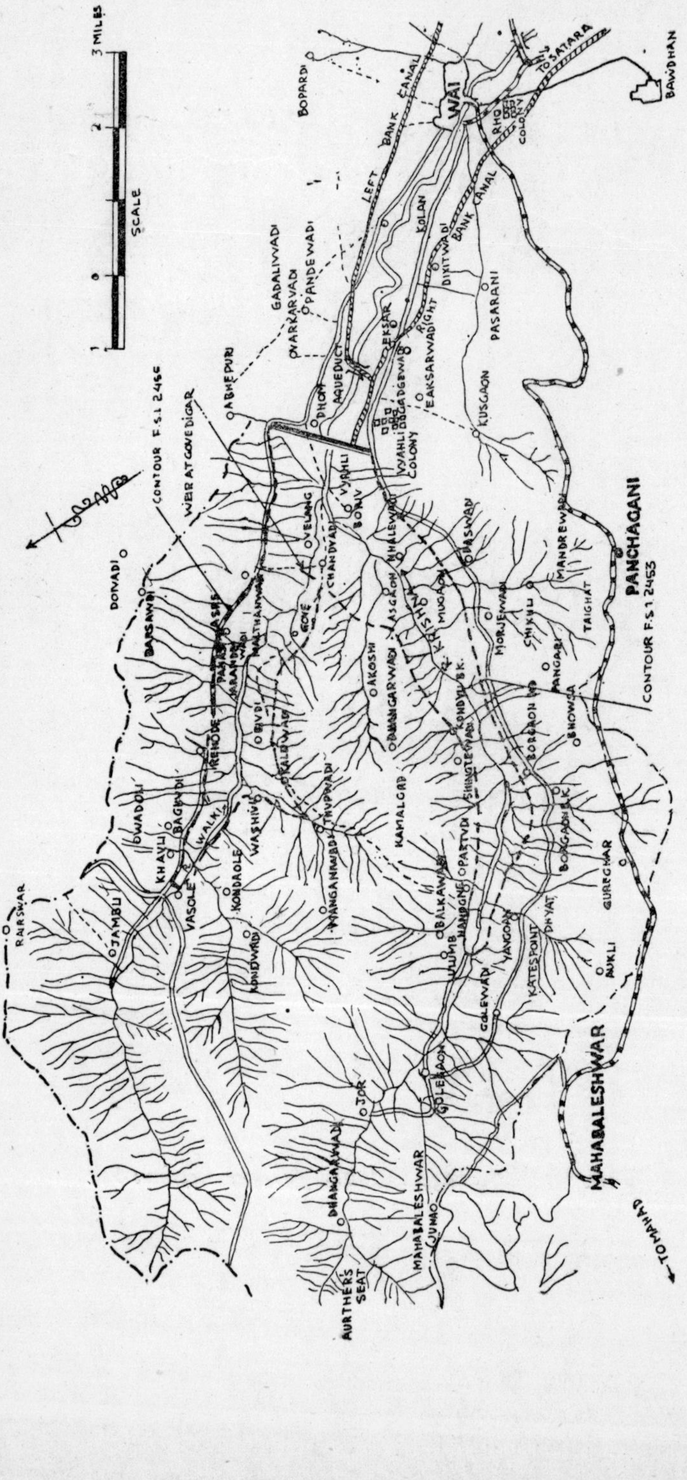

FIG. 3.3 : *Drainage Pattern of Catchment Area of Dhom Dam (area under submergence)*

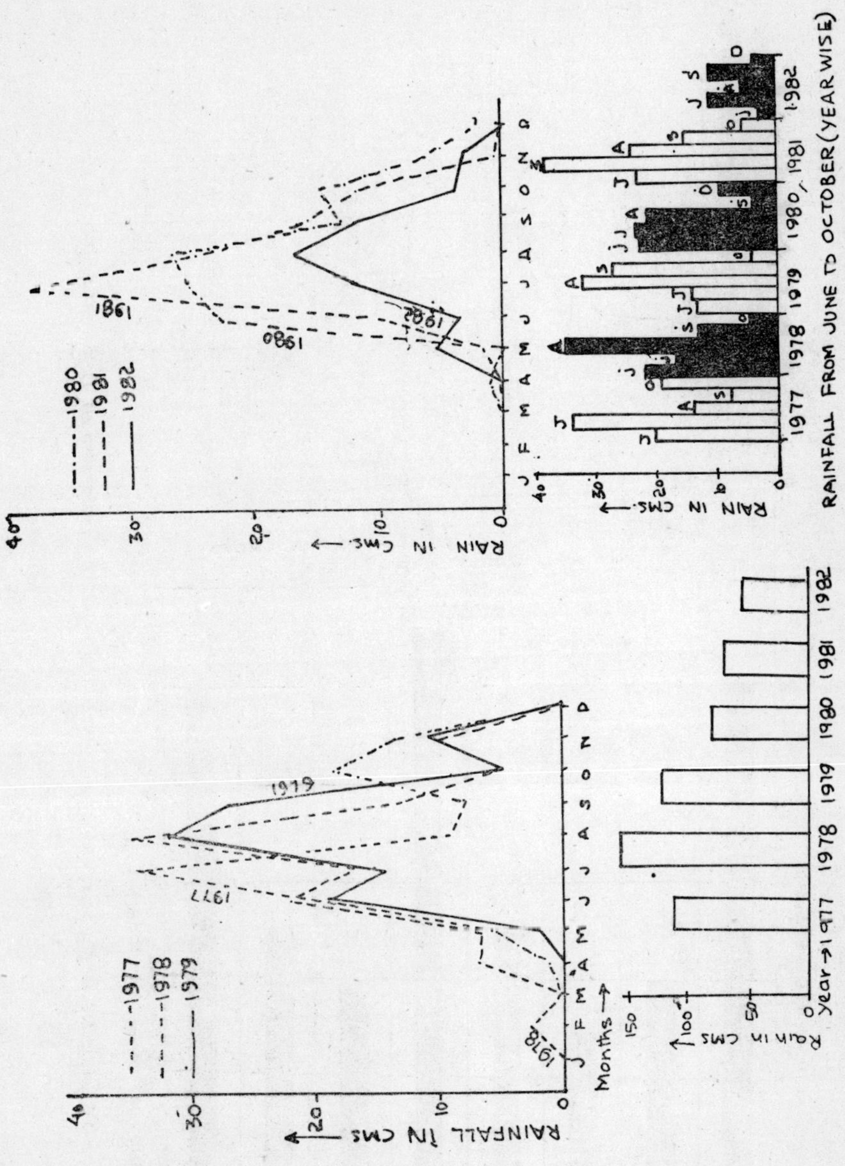

FIG. 3.4 : *Bar and Line Graph—Rainfall of Village Pasarani (at the vicinity of Dhom Project)*

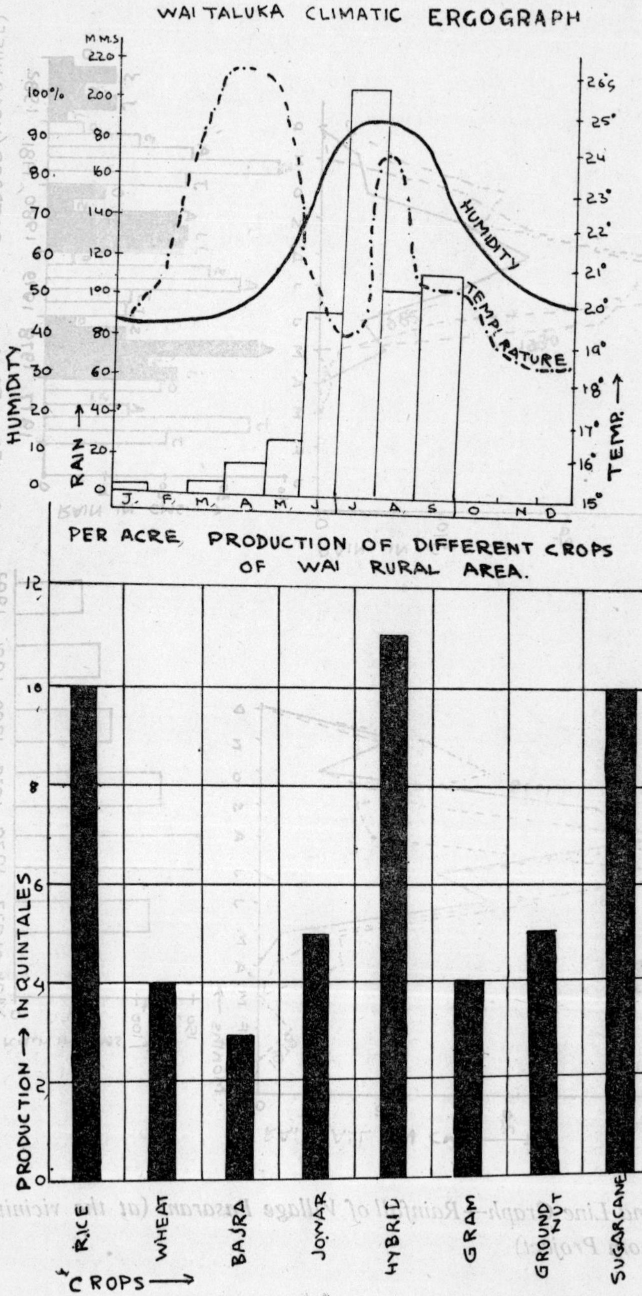

FIG. 3.5 : *Climatic Ergograph of the Region and Per Acre Production of Different Crops of the Area*

receiving 25" and less rainfall (Fig. 3.6). Now this zone is under the canal benefited area of Dhom dam.

Vegetation

The western part of the tahsil, together with its hill slopes and spurs reaching the river course is largely under forest. There are teak-sag (Tectona grandis), Anjan (Hardiwickid binata), Behada (Fermindalia belerica), Chinch (Termarindus indica), Hirda (Fermination chebula), Chandan (Santalum album), Karanj (Pongamia glabra), Khair (Acacia Catechu), Nimb (Melia Azadirachta), Mango (Mangifera indica), Umbar (Ficus glomrata), Vad (Ficus bengalensis), Pimprani (Fiew Arnottiana), Babul (Acacia Arabica), Jambul trees. But the forest cover is diminishing slowly both in area and density of trees, and the slopes are now covered with rough pastures, scattered trees and open scrub. On the slopes, thickets of Karvanda (Carssa-carandas) is common.

Babul (Acacia Arabica), Bor (Zizyphus Junjuba), Nirgudi (Vitex Negundo), Ghayapat (Agave Sisalane) Karvand (Carssa-Carandas), Nivdung (Fuphoriba nerifolia), Nagphani (Opuntia Dilleni) are found in the eastern zone. Many hill faces are simply bare rock exposures.

In the past, the region was completely under dense forest area and it was very difficult to travel alone from one village to another along the Krishna valley and in the adjoining hills. The villages of Pasarni, Baleghar, Lohare, Abhepuri. (Fig. 3.0) before independence, were well-known for wild pig hunting in the forest areas. Bones, Barcha (a weapon) were stored by one of the families as a memory of wild pig hunting of the village Lohare.

Medicinal plants in the area of Baleghar and Pandavgad were also well known. Lumbering for collection of gum, honey medicinal herbs and traditional occupations; and a recent addition is sericulture based on mulberry.

Main Aspects of Agriculture

Broadly in Wai taluka, the area below the contour line 788.68 m. is a fertile area (Fig. 3.1), largely under cultivation of sugarcane, turmeric and vegetable production. Sugarcane is an important commercial crop and its cultivated area under canal irrigation is increasing rapidly within the whole region. The main food crop of the dry zone is jowar and of the wet zone is rice, wheat, bajra, nachani, groundnut are the other crops, but the area under these crops has declined due to greater importance of irrigated crops. Jowar, groundnut and wheat are now giving way to commercial crops like sugarcane, turmeric and vegetables, specially along the sides of the canals.

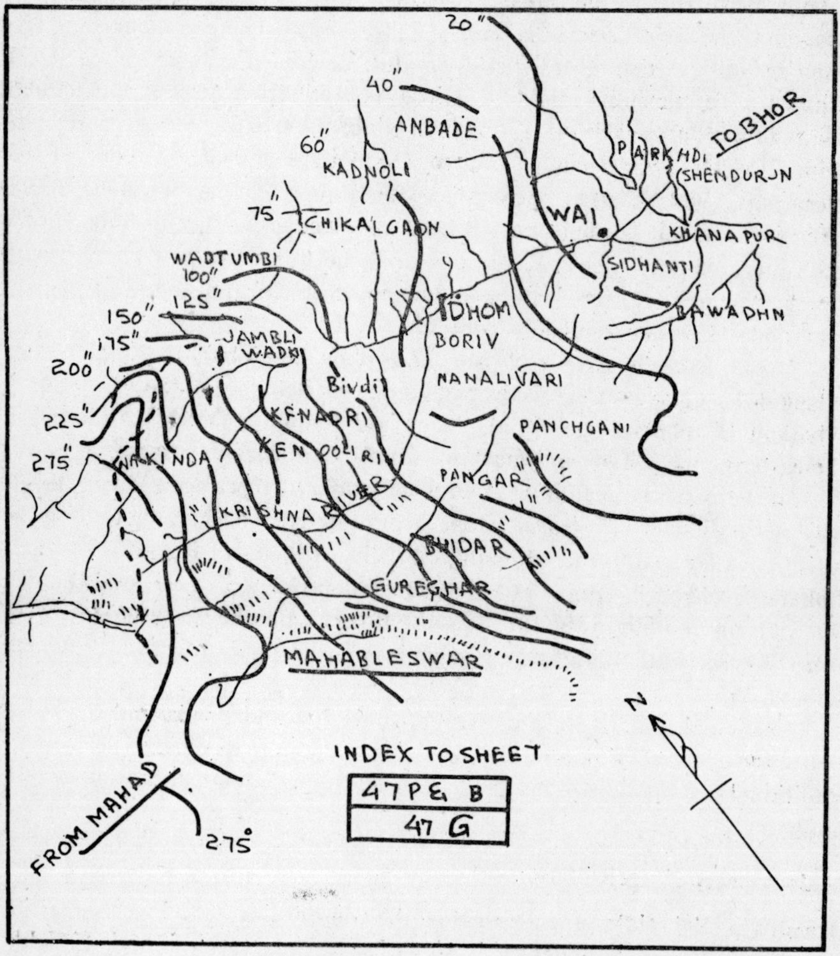

FIG. 3.6 : *Isohytes in the Catchment Area of Dhom Project (Rainfall in Inches)*

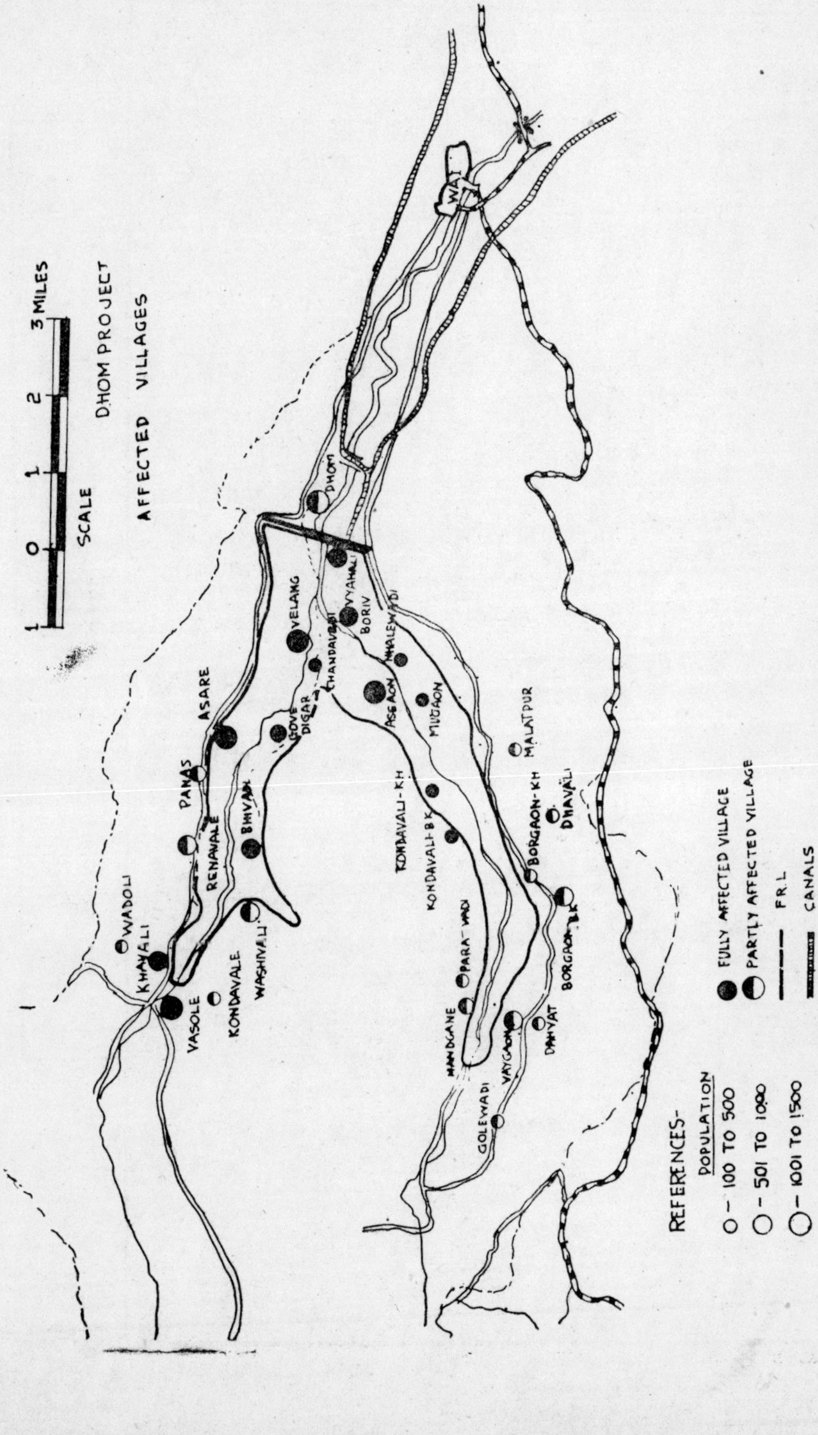

FIG. 3.7 : *Affected Villages by Dhom Project*

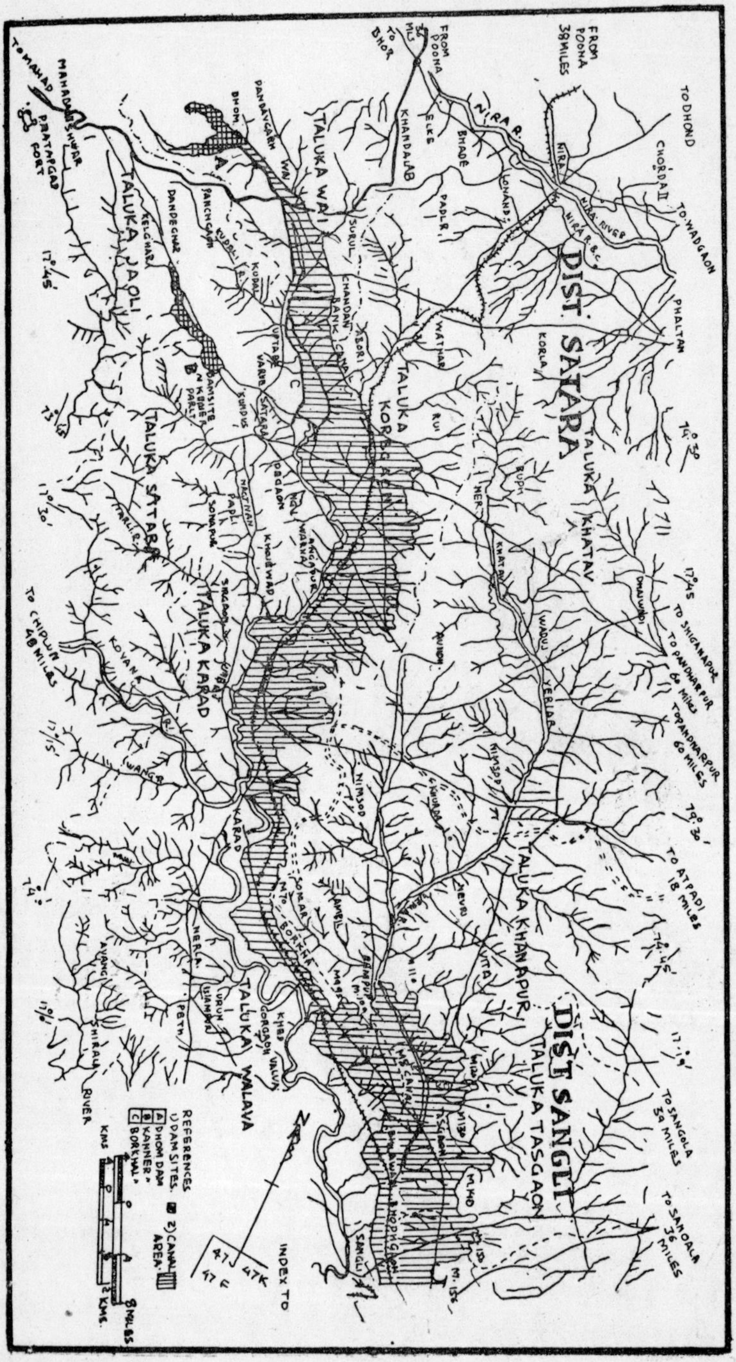

FIG. 3.8 : *Krishna Dhom Project—Command Area with Canals*

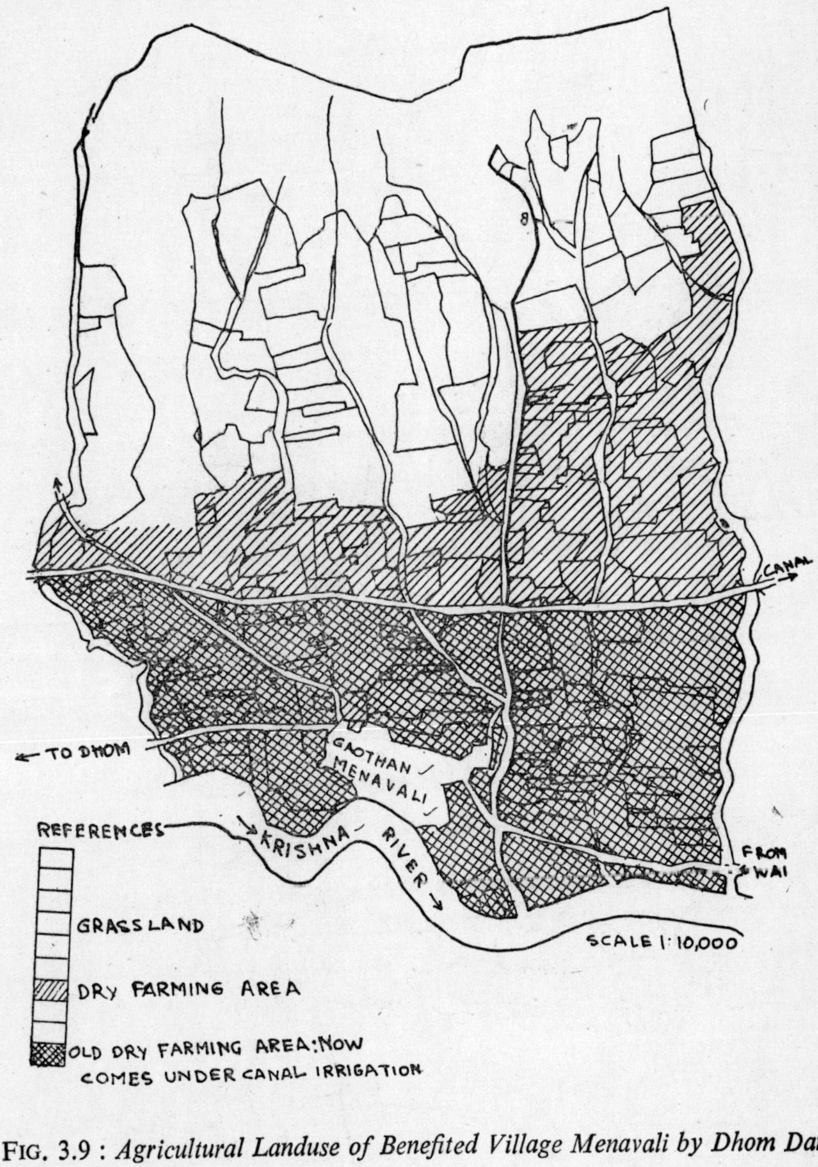

FIG. 3.9 : *Agricultural Landuse of Benefited Village Menavali by Dhom Dam*

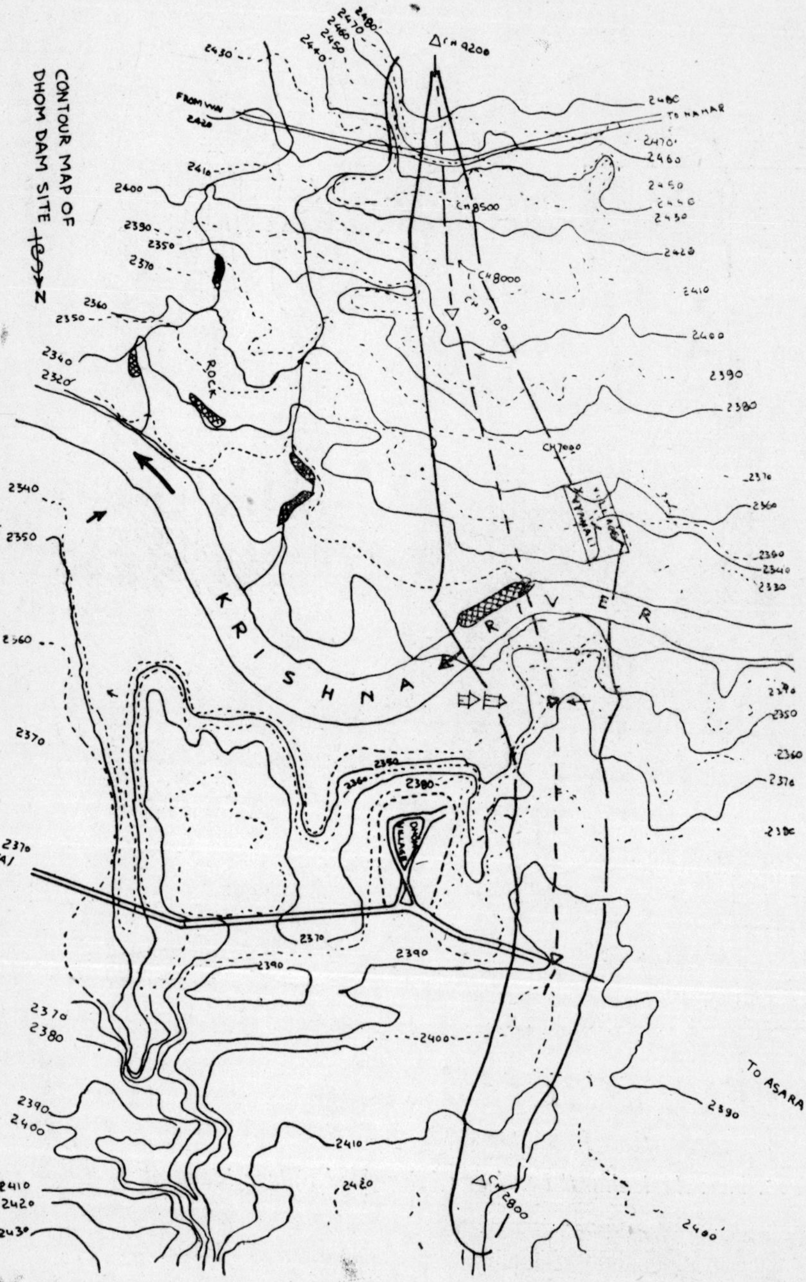

FIG. 3.10 : *Dhom Dam Site—Contour Map and the Wall of the Dam*

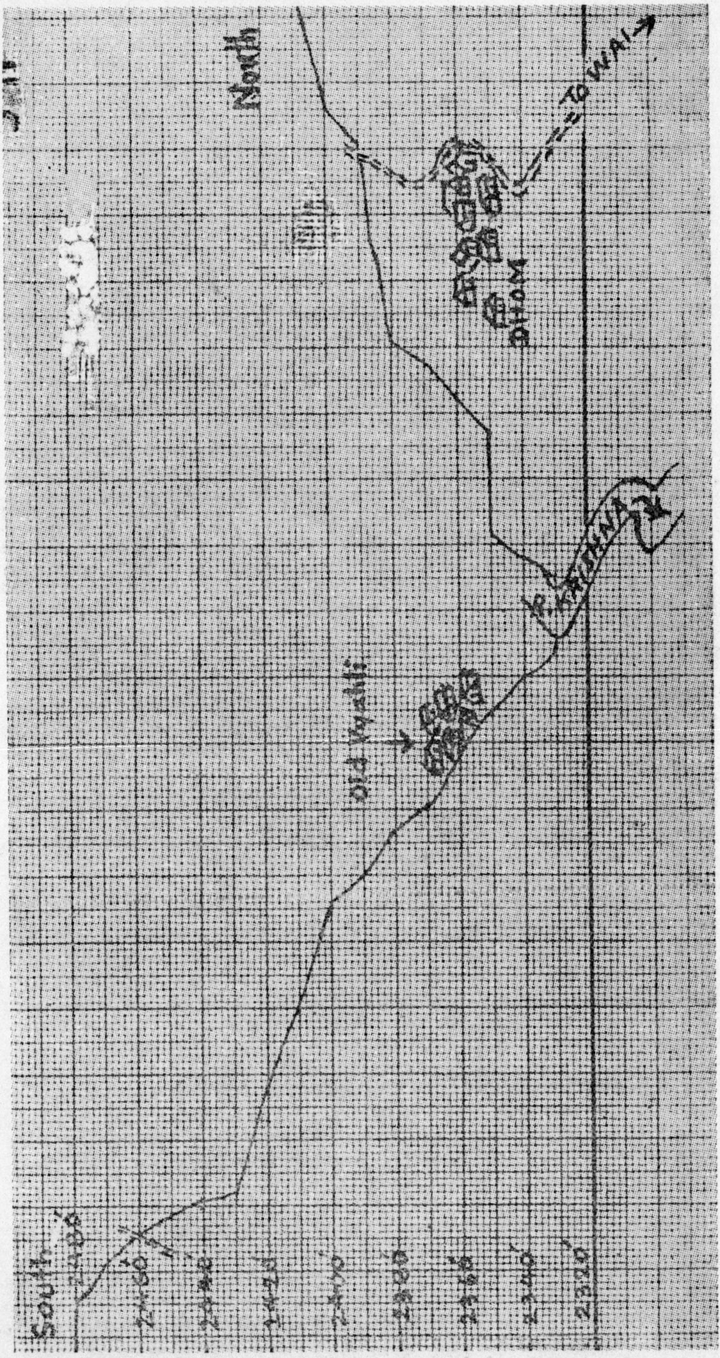

FIG. 3.11 : *Cross-Section of Dhom Dam*

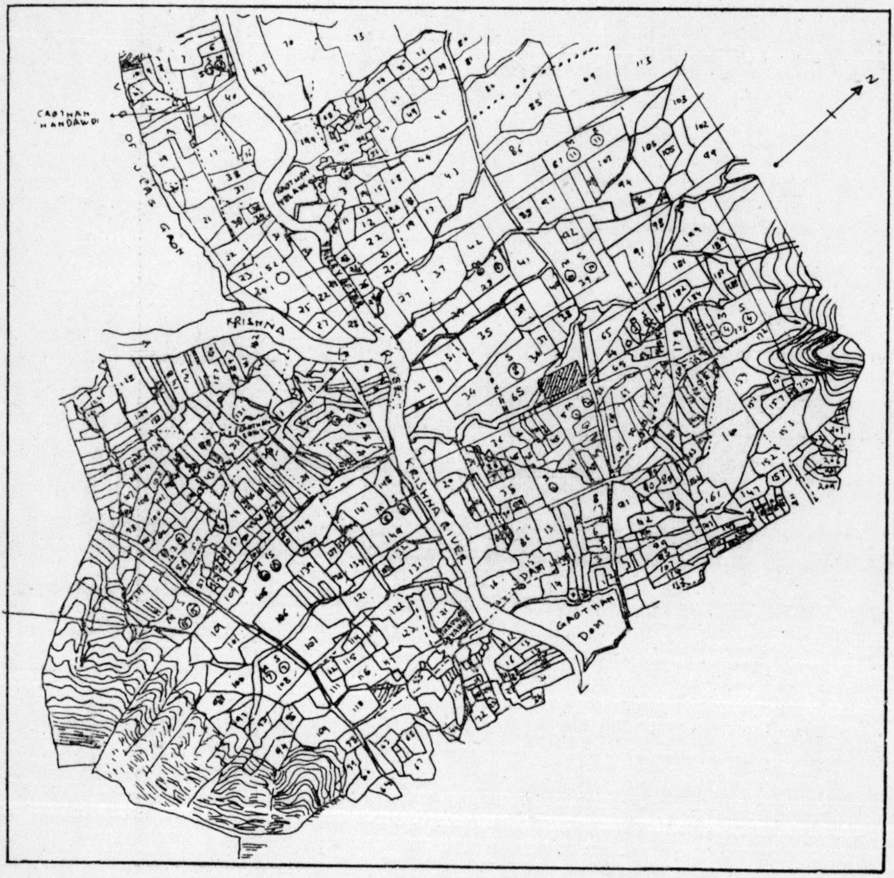

FIG. 3.12 : *Borrow Area for Dhom Dam*

TABLE 3.1 : **Growth of Area Under Irrigation (in hectares) in Satara District, 1973-74 and 1981-82**

	1973-74	1981-82
Government canals	—	6520
Private canals	2217	2217
Wells	3284	3284
Other sources	—	—

Total net irrigated : 5501 ha.

Total gross area irrigated : 7286 ha.

Other Features

Farming is, by and large, the traditional occupation. The villages of the western zone are very backward and traditional. The natural environment is quite pleasant. But no proper roads are maintained. People are also engaged in other primary types of occupations (i.e., hunting, firewood cutting, collecting fruits like Jambhul and Carssa-carandas).

The eastern zone was earlier known for its dry farming. Now due to canal irrigation, the area becomes very important for irrigated crops.

Along the Pune-Bangalore highway, the prosperity of the eastern zone is slowly growing under sugarcane cultivation, and the sugar factory at Bhuinj is its visible expression. (The Satara Sahakari Sakhar Karkhana Ltd., Bhuinj, Tal. Wai). This is the region, where the new locations of dam affected villages like Vyahali and Chandavadi are located. Bhuinj and Panchwad are important urban centres.

In future, the surrounding settlements of the command area are expected to become urbanized locations and already land values along the Wai Dhom belt are increasing. The occupational structure is becoming firmly agro-based.

DHOM PROJECT AND VILLAGE DHOM

Technical aspects of Dhom dam

Village Dhom (73° 49′ E Long and 17° 59′ Lati) situated on the left bank of river Krishna at a height of 716.28 m. A.S.L. It is connected by a 9 km. unmetalled road with Wai. (Fig. 3.7). Dhom is 5 kms. away from district place Satara. The dam construction affects the western part of Dhom village (Fig. 3.10). Nearly, one-third of the geographical area of Dhom has gone under the dam itself.

The location of the Dhom dam proper is to the west of Dhom village at a distance of 93 metres. The length of the dam is 2478 metres, and the

TABLE 3.2 : **Dhom Dam—Canal Irrigated Area and Benefited Talukas by Canal Irrigation**

Canal	Length (in kms.)	Irrigated area (in Hectares)	Benefited Talukas
Right Bank Canal	44	6192	Wai, Jaoli, Satara
Left Bank Canal	93	19160	Wai, Satara, Koregaon

Taluka-wise Number of Villages Benefited by the Krishna Dhom Irrigation

Taluka	Wai	Jaoli	Satara	Koregaon	Total
No. of benefited villages	32	7	13	20	72

maximum height is 50 metres (Fig. 3.10). Area occupied by Dhom dam is 2400 hectares.

The capacity of the water reservoir is 255.18 million cubic metres. The total land irrigated by the project is 217.56 sq. kms. There are 29 affected villages by Dhom dam. (Fig. 3.7 and 4.0)

In September 1982, the water stored in Dhom dam was about 51 million cubic metres. It has a capacity to store 353.56 million cubic metres of water. According to the government booklet published on 30th March, 1969, the capacity to store water by the Dhom lake is estimated at 352 million cubic metres. Capacity of Koyna dam is 9804 T.H.C. and upto September 1982, water stored in it was only 72.2 T.M.C.

The construction work of the Dhom dam was started in March 1969, and was completed in 1977. Canal construction work was completed in 1981. Thus nearly 9 years period was taken to complete the construction work of the Dhom project.

From Panchagani, it is worthwhile to see the beauty of "Dhom Lake", a man made resource for economic development of the region. There is now a beautiful garden near Dhom project maintained by the Irrigation Department.

On 16 September, 1982, at 8.00 P.M., the level of the water in the dam was 745.70 m. (Aug. 81) and 748.30 m. (Aug. 82) and the storage capacity is 382.32 million cubic metres.

The Physical Setting and Future Prospects

Thus it will be seen that in the upper Krishna valley we have a region of marked physical and climatic contrasts, and correspondingly a markedly varied human response. In the western part of rugged hills dependence on deteriorating forests and poor farming are very characteristic and in the eastern comparatively flatter and decidedly drier part the regional community, by and large, lives on poor agricultural practices, under uncertain rainfall conditions. Yet this physical diversity has a unity created by the Krishna river which develops good soils on its banks and those of its tributaries, attracts human settlements and facilitates their social and economic contacts by foot-paths and roads. The ancient town of Wai attracts people but more powerful is the pull of the hill stations of Panchgani and Mahabaleshwar overlooking the Krishna valley. What is provided by nature can be certainly strengthened by human technology and wisdom. It is in this context that the Dhom project becomes important not only for the rehabilitation of the affected but for balanced regional development as a whole – with well preserved scenic beauty in the hills and prosperous community in the well tilled valley supported by canal irrigation. The prospects of reviving old historic place of sanctity Dhom into an attractive tourist centre and renovating the other holy place of Wai into a holiday home for the aged and the pious and for developing dairying and other agro-based industries are bright and the regulated flow of the river would make Wai one of the most pleasant towns in Maharashtra.

4

The Affected Villages

The Old and the New Sites

Introduction

IT IS remarkable, that due to Dhom project, as per census record, the total population of Wai Tahsil has been reduced by 6,722 population.

1971—Population of Wai Tahsil=125099
1981—Population of Wai Tahsil=118377
Project affected and shifted population outside the Wai Tahsil=6722

The total population of the 29 affected villages was 17784 and the total area covered by them was 2049.8 ha. (5704 acres) including the village site and agricultural land (Fig. 4.0). Out of the 29 villages, 14 were completely affected and 15 partly affected villages by the project. Table 4.1 gives details.

Decrease of Population of Fully Affected Villages

As compared the population data with 1971, all the affected villages are observed that their population is decreased in 1981, due to shifting of population from the old to the new locations (Table 4.1). The data of 1981 population is indicating remained population at old sites, settled at the protected upward part from the water level of the Dhom lake, though they are included as fully affected villages being loss of 75 per cent of agricultural and settlement area, under the project. The population of six villages [Asare, Vasole, Velang (including Umberwadi), Vyahali, Kondavale Bk. and Khavali] is in between 400 to 1,100, which is remained settled at old location and their 30 to 50 per cent of population is shifted elsewhere.

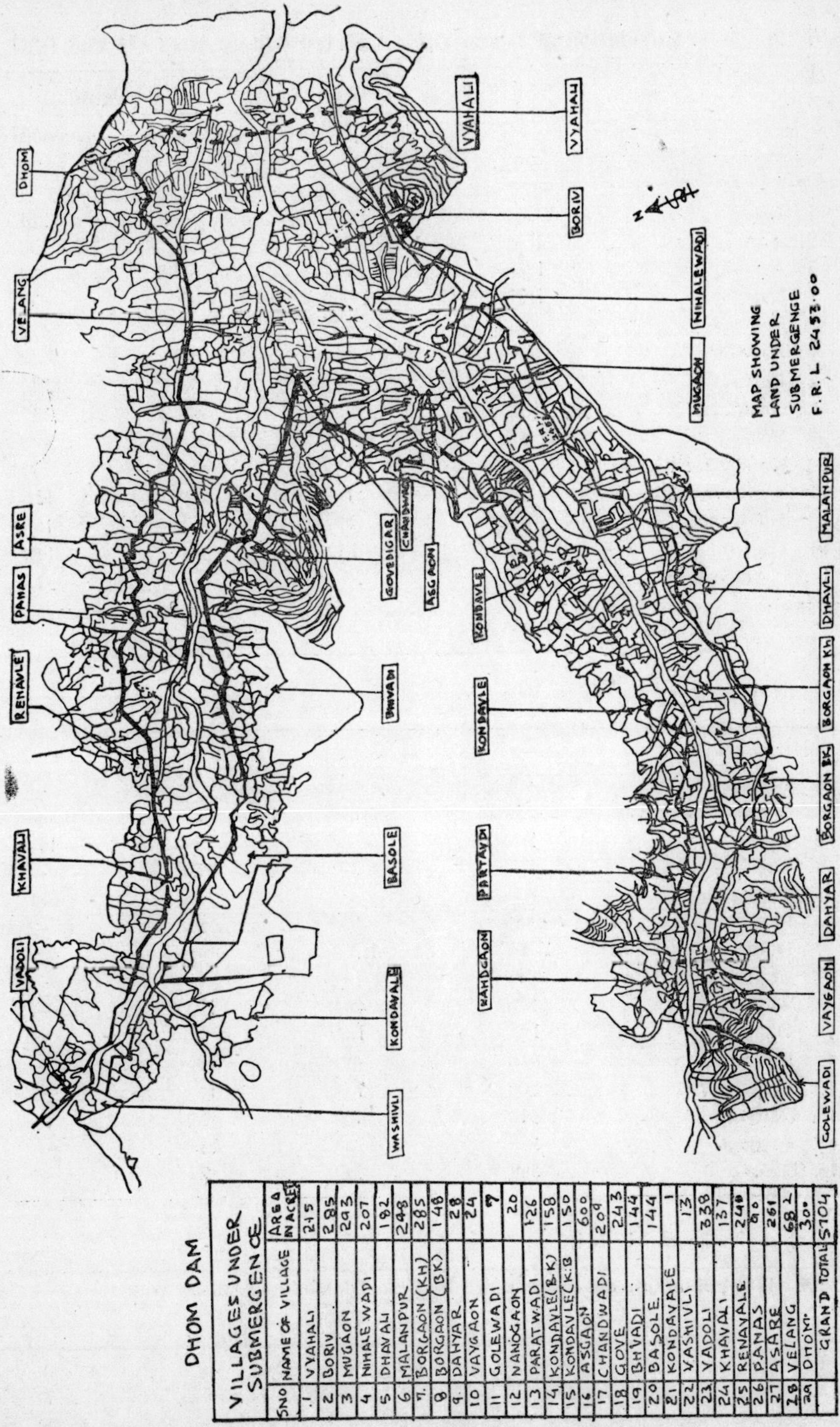

DHOM DAM

VILLAGES UNDER SUBMERGENCE

SNO	NAME OF VILLAGE	AREA IN ACRES
1	VYAHALI	215
2	BORIV	285
3	MUGAON	293
4	NIHALE WADI	207
5	DHAVALI	182
6	MALANPUR	248
7.	BORGAON (KH)	285
8	BORGAON (BK)	148
9.	DAHYAR	28
10	VAYGAON	24
11	GOLEWADI	7
12	NANOGAON	20
13	PARATWADI	126
14	KONDAVLE(B.K)	158
15	KONDAVLE(K.B	150
16	ASGAON	600
17	CHANDWADI	209
18	GOVE	243
19	BHVADI	144
20	BASOLE	144
21	KONDAVALE	1
22	VASHIVLI	13
23	VADOLI	338
24	KHAVALI	137
25	RENAVALE	248
26	PANAS	90
27	ASARE	261
28	VELANG	682
29	DHOM	30
	GRAND TOTAL	5704

FIG. 4.0 : *Land Under Submergence of Dhom Dam*

TABLE 4.1 : **Population of Dhom Dam Affected Villages in 1971 and 1981**

	Population of 1971	Population of 1981	Population	
			Increased	Decreased
Fully affected villages				
1. Asare	1,415	1,054	—	361
2. Velang	1,357	923	—	434
3. Asgaon	1,231	127	—	1,104
4. Vasole	1,075	1,036	—	39
5. Vyahali	937	475	—	462
6. Khavali	634	718	84	—
7. Bhivadi	608	65	—	543
8. Govedigar	585	68	—	517
9. Boriv	567	121	—	446
10. Kondavale Bk.	542	472	—	70
11. Kondavale Kh.	284	168	—	116
12. Nhalevadi	377	192	—	185
13. Mugaon	322	152	—	170
14. Chadavadi	314	—	—	314
Total	10,245			
Partly affected villages				
1. Partavadi	459	482	23	—
2. Nandagane	322	344	22	—
3. Panas	202	48	—	154
4. Vadoli	283	307	24	—
5. Malatpur	441	440	—	1
6. Renawale	889	955	66	—
7. Dhom	1,367	1,285	—	82
8. Kondavale	499	—	—	—
9. Washivali	540	645	105	—
10. Dhavali	175	185	10	—
11. Borgaon Kh.	460	495	35	—
12. Borgaon Bk.	566	734	168	—
13. Dahyat	413	463	50	—
14. Vaygaon	579	762	183	—
15. Golewadi	344	366	22	—
	7,539			
Total of 29 villages=17,784				

Note : 1981 population data is obtained from the Mamalatdar Office, Wai.

These villages are included under "fully affected category" due to their great loss of agricultural land than the settlement area under the project. The population of seven villages, (Table 4.1 data 1981 population), Asagaon, Bhivadi, Govedigar, Boriv, Kondavale Kh., Nhalewadi, Mugaon, is in between 100 to 200 indicates that very few people are settled at old location. These villages have lost their major agricultural and settlement land under the project. (See definition of "dam affected person" Chapter 5). Chandavadi is the village, which comes under no population, agricultural and settlement land is remained at old location. This village is completely shifted at its new site, at Khadki village area.

Increase of Population of Partly Affected Villages

From Table 4.1, the rate of increased population, of partly affected villages is very small, which indicates as it is explained by some partly affected villages, that they were under very high pressure of "shifting problem of their village by the project", and started to migrate from 1976 and onwards at nearby urban centres for obtaining employment and at the villages of their relatives to settle their families. Population of all the partly affected villages is below 1,300 and all the villages, except Panas, are remained at old locations and their major agricultural land is acquired by the project.

Pattern of Location

It will be seen that there is a pattern of location in these villages (Fig 4.2) and Fig. 3.7 The bigger villages were well situated in the valley courses of the Krishna and its tributary Walki, while the small villages were well away from the course and situated at the junction of valley and hill slopes.

Locational Characteristics

The affected villages, those which lost their village sites as well as their land, were both, small and large villages. The large villages were Dhom, Boriv, Velang, Vyahali, Asagaon, Asare, Velang were functioned as a service centre of the area. Agriculture was the main occupation of these people with 80 per cent working directly on the land. However their subsidiary occupations were also important. Asare, Asagaon, Govedigar, Bhivadi, Kondavale Bk. and Kh., were connected with their nearest market centre - Panchagani—a hill station of Mahabaleshwar Tahsil which provided to them. Subsidiary occupations, like marketing of thatched roofs for buildings to protect them from heavy monsoon rain; and some of the villagers used to work in the summer season as "Malis" in private gardens; poultry keeping, dairying and collecting forest materials (i.e., Jambhul, honey, gum, wood, phanas, mango, etc.) were other subsidiary occupations. Wai Taluka place was another market centre to all the affected villages. Their reaction is more severe due to the total loss of village site and its agricultural land. They felt uprooted. They could not accept the reality that their ancestral homes and social environment had literally gone forever under the impounded lake waters.

Partly Affected Villages

There were two kinds of partly affected villages:

(*i*) Those which partly lost their land; and (*ii*) those which partly lost their village sites as well as some land.

Some of these villages are situated not at the brink of the impounded waters and others also well away from the waters but were liable to seasonal. The residents of the partly affected villages now live at upper levels well away from their old sites. The residents, however, feel that these new settlements are quite unlike old crowded ones. They feel a strangeness about their new site, closeness of the impounded waters is a new experience to them and the memory of their old village lingers. Their occupations are almost the same as they were before, and the possibilities of new agro-based activities like dairying, cattle breeding, irrigated cash crops are still to be explored by them.

From 15 partly affected villages, 20 per cent sample families were studied and it is found that those who have lost their agricultural land, and saved settlements from the project, are observed very uneasy due to difficulties for cultivating small patches of agricultural land obtained in command area, which is far away from their settlement locations at different places and that too 8 to 10 kms. away. To manage the farming operations is very tiresome work to all of them. They cannot keep contact with their villagers and they cannot enjoy their social village life due to much of time is invested for maintaining and visiting their agricultural farms. They lose their village unity. Such families have started to move and shift to the agricultural farms by constructing huts as a shelter for residential purpose within the farms, and slowly they were removed away from the village social life. Such huts are very necessary for them, for taking care of their cultivated crops. The families who have lost their settlement area under the Dhom project, but saved their agricultural farms which comes under infertile mountain slopes, their remained unaffected land is only useful for grazing purpose. Cattle rearing and cultivation of rainfed crops are the only economic sources remained to them. It is possible for the government to provide the upward mountain slopes to them for grazing purposes. It is very expensive to the government for providing water for agricultural purposes by sanctioning "lift irrigation schemes" from the stored water of the "Dhom lake", to the agricultural land of all the partly affected villages.

The Analysis Based on Population Characteristics and Infrastructural Facilities as Existed in the Year 1971

There are in all 14 villages which are submerged and fully affected due to the construction of Dhom dam, which began in the year 1969. The analysis is based on population characteristics and the infrastructural facilities as existed in the year 1971.

Population Size

Among the 14 villages the population size ranges from 284 in Kondavale Kh. to 1,415 in Asare village. There are 4 villages having population in the range of 1,000 to 1,500; 6 villages in the range of 500-1,000; and the remaining 4 villages between 250 to 500 and the largest village submerged was on the bank of River Walki.

Rural Area

In terms of village over the acreage ranges from 201 acres of Chandavadi (Closer to the confluence of rivers Walki and Krishna) to 2,453 acres of Vasole. Majority of the villages (10 villages) have the rural area between 500 to 1,000 acres. Only two villages have rural area between 1,000 to 2,000 acres and only one village having less than 300 acres.

Density of Population

The average density of population in the submerged villages was almost one person per acre. The per person acreage is highest (2.25) in Vasole and lowest (0.35) in Asare which also indicate the relative low density of population and high density of population respectively. The per capita area available was 1.20 acres in these villages. The majority of the villages (12) are within one Standard Deviation (+ or –) and remaining two villages Chandavadi and Asare showed higher density of population.

Households to Houses

The ratio of households to houses is calculated to know the existing condition of housing facility. Out of 14 villages only 4 villages (Asare, Asgaon, Kondavale Bk., Kondavale Kh., are adequate number of houses to accommodate their households. The remaining villages show an inadequate housing facility *Low inadequacy* in 4 villages (Vyahali, Velang, Vasole and Khavali); *Moderate inadequacy* in 5 villages (Nhalevadi, Bhivadi, Govedigar, Chandavadi and Boriv); *High inadequacy* in only one village—Mugaon. Except one (Mugaon) all other villages show variation of one Standard Deviation (+ or –).

Literacy

There is a greater variation in the percentage of female literacy than in the male literacy of these villages. The percentage of female literacy varies from 3 in Bhivadi village to 18 in Boriv village whereas male literacy varies from 22 per cent in Khavali to 54 per cent in Nhalewadi. By and large the villages have 30 to 45 per cent of literate male population (12 out of 14 villages) and 5 to 15 per cent of literate female population (8 out of 14 villages). Govedigar and Mugaon show higher dependence.

Dependency Ratio

The dependency ratio ranges from 1.08 in Bhivadi and Kondavale Bk. to 4.47 in Asagaon which indicate very few dependents and too many dependents respectively. The high dependency ratio in Asagaon is mainly because of the very low female participation in work. Similarly in Mugaon the dependency ratio is 3.24 and there is no female participation in work. In general, the villages having low female participation in work show relatively high dependency ratio.

Workers in Agricultural Sector

Except Vyahali village, other villages have more than 75 per cent of their workers engaged in agricultural sector. The more predominant agricultural activity is in Bhivadi (99%), Asare (97%), Kondavale Bk. and Kondavale Kh. (95%), Khavali and Vasole (94%), in Vyahali only 35 per cent of its workers are engaged in agricultural activity. This is perhaps because of its closer location to dam site which was affected first in terms of changes in economic activities. The workers were attracted to the dam site for earning their livelihood.

Workers in Secondary Sector

The proportion of workers engaged in secondary activities vary from less than one per cent to nearly 20 per cent. Vyahali, Chandavadi and Govedigar which are closer to the dam site and respectively have higher proportion of workers in secondary sector and more so in construction activity (19.78%, 14.18% and 13.79%). The population of workers engaged in quarrying activity is also significant at Vyahali (Closer to dam site). The proportion of workers engaged in secondary activity forming less than 1 per cent is observed in Khavali, Asare and Bhivadi which are located away from the dam site. From the map it is observed that the villages closer to the dam site, such as Vyahali (19.78%), Chandavadi (14.18%) Govedigar (13.79%), Nhalewadi (11.31%), Asagaon (10.22%), Boriv (9.70%), Mugaon (9.21%) and Velang (7.05%) have relatively higher proportion of workers engaged in secondary activities.

Predominance of Female Population

The predominance of female population is observed in all villages except one, i.e., Mugaon. 9 out of 14 villages have female proportion exceeding 1,200 per 1,000 males. Among these villages two villages namely Nhalewadi and Asare show more than 1,300 females per 1,000 males. This phenomenon is partly because of initiation of dam construction activity itself which has attracted menfolk to the dam site, leaving behind their women-folk to look after the agricultural fields (may be an area of heavy out-migration).

Female Participation in Work

The female participation in economic activity is calculated to the total female population. The proportion ranges from zero (Mugaon) and nearly zero (Asagaon 0.15%) to nearly 57 per cent at Bhivadi 4 out of 14 villages—Bhivadi, Kondavale Kh., Kondavale Bk., and Chandavadi show greater female participation in economic activities (more than 50% of its female population). Another 4 villages have 40.50 per cent of its female folk engaged in economic activities. The proportion of female workers in 9 out of 14 villages is more than 50 of its total folk force.

Social Amenities

In the submerged area there were 14 villages located on the banks of Walki and Krishna rivers. The river water was available to all the villagers for domestic use. In addition to the river water, kutcha road facility and primary school facility was also available to all the villages. The facility of electricity was available only to those villages which were closer to the dam site (Velang, Asagaon, Vyahali, Boriv, Nhalevadi and Chandavadi). Only 5 villages had the facility of Post Office of which 4 villages were longer (having 1,000 population and more). Asagaon had middle school, High School and Health Centre and Velang had Middle School. In terms of functional ranks these two villages emerge as first and second respectively. These are also respectively located in Krishna basin and Walki basin.

SOME CHARACTERISTICS OF AFFECTED VILLAGES

Population and Cultivators

Asare, Dhom, Velang. Asagaon, Vasole, Vyahali and Renawale, ranks within the population of 900 to 1,500. All the villages were located on gentle slopes and on extensive fertile plain area of the lower part of the Valley. Agriculture (Rice cultivation) and forest products, were the main income source to them. The same villages were covered within 200 to 500 cultivators due to the availability of extensive agricultural land.

Total Geographical Land : Irrigated and Unirrigated Land

Due to availability of mountainous rugged topography of Kondavale, Vasole, Velang, Dhom, Asagaon, Golewadi and Khavali, were the major

TABLE 4.2 : **Dam Affected Villages—Area Under Submergence of Dhom Dam**

Name of the villages	Total area of old village in acres	Area under submergence of Dhom Dam in acres	Percentage of affected submerged area of the old village
Fully affected villages			
1. Asare	494	261	52.83
2. Asgaon	1,261	600	47.58
3. Velang	1,669	682	40.86
4. Vyahali	887	215	24.23
5. Chandavadi	201	201	100.00
6. Boriv	466	285	61.15
7. Govedigar	666	243	36.48
8. Mugaon	527	293	55.59
9. Bhivadi	651	144	22.11
10. Khavali	969	137	14.13
11. Vasole	2,453	144	5.87
12. Kondavali Kh.	581	158	27.19
13. Kondavali Bk.	913	150	16.42
14. Nhalewadi	548	207	37.77
Total	12,286	3,720	30.77
Partly affected villages			
1. Partawadi	756	126	16.66
2. Nandagane	661	20	3.02
3. Panas	127	90	70.86
4. Vadoli	601	338	56.23
5. Malatpur	712	248	34.83
6. Renawale	947	249	26.29
7. Dhom	1,263	300	23.70
8. Kondavale	2,643	1	0.03
9. Washiwadi	882	13	1.47
10. Dhavali	271	182	67.15
11. Borgaon Kh.	565	285	50.26
12. Borgaon Bk.	745	148	19.86
13. Dahyat	614	28	4.56
14. Vaygaon	736	24	4.56
15. Golewadi	1,135	7	0.61
Total	12,658	2,059	16.26
Total of all the villages	24,944	5,779	23.16

villages—with more than 1,000 acres of grazing land along their nearby mountain slopes. Along the bank of the river Krishna and Walki, river irrigated land was in between 100 to 200 acres and oil engines were used for irrigation purpose by such villages. Well irrigation was also important source to them. There were no electrified wells, but bullocks were the main power utilised for well irrigation of the land. There were Velang, Wasole, Dhom, Asare, Vyahali, Mugaon, and Kondavali villages ranks within 300 to 700 acres of unirrigated land, under the grasslands on the upward mountain slopes, which is lost under the project. The villages having major forest land were covered with westward part and lower part of the valley. Westward part of valley comes under wet climatic region with high rainfall.

All the affected villages (total 29 villages), have lost 23.16 per cent of their land out of their total geographical area under the dam submergence (5,779 acres out of 24,944 acres) (Fig. 4.0 and Table 4.2).

The maximum water level of Dhom Lake is shown in Fig. 4.0 by a line showing F.R.L. (Flood Reduced Level) 2,453′. Rice, ragi, jowar were the staple food crops of the valley. To the total land, irrigated land was only 13.5 per cent and wells and river were the main irrigation sources for agriculture. Educated persons were only 21.6 per cent.

Village Dhom – At its Old Location

Population in 1971 was 1,367. In 1981 was 1,285. For the present, still two-third of the population of Dhom is unshifted and settled within the old location (just below to the left side of the dam wall, on the left bank of river Krishna), up to this time and is in a suspense. It is afraid that the government will at any time order them to leave the village and to shift at Joshi Vihir. At new location in the village area of Ozarde (Tal. Wai). Some of them have obtained agricultural land in Bhuinj village area. Except 3 to 4 plots, no developments has taken place in the new location. At the old location the villagers are unwilling to repair or construct buildings and roads. To the north of this old location, there is a mountain range with a maximum height of 728.48 m. (2,394′) above sea level extending east-west direction with a boundary of village Abhepuri and to the east is the boundary of village Bhogaon (a village of well-known Marathi poet of 17th century Waman Pandit Samadhi). The southern boundary of Dhom is limited by River Krishna, there are many Hindu temples. Mahadeo-Dhomeshwar Temple is the chief temple. The arches and inside walls are of polished basalt. Mahadeo Ling of this temple, is said to have been set up by a Rushi hermit named Dhoumya and hence the village is named Dhom.

Area covered by village is 17 acres with compact and irregular settlement

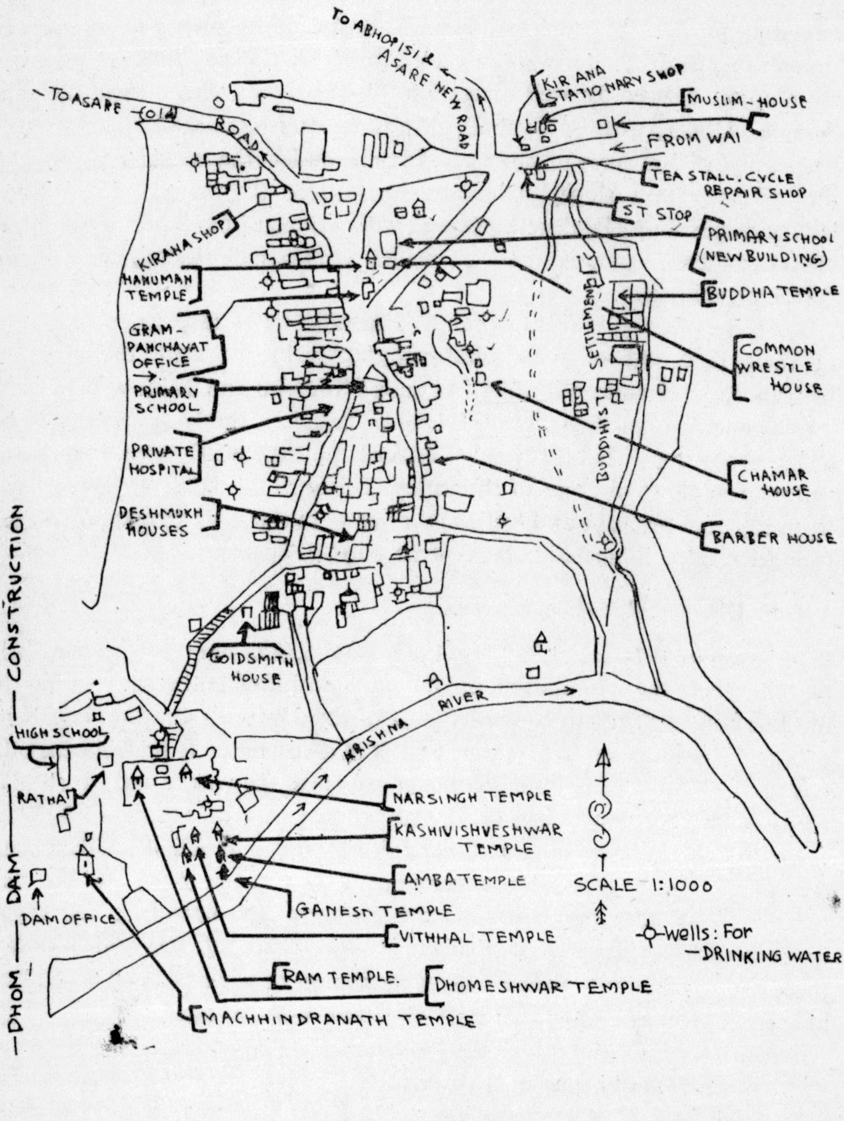

FIG. 4.1a : *Sketch Map of Village Dhom at Old Location*

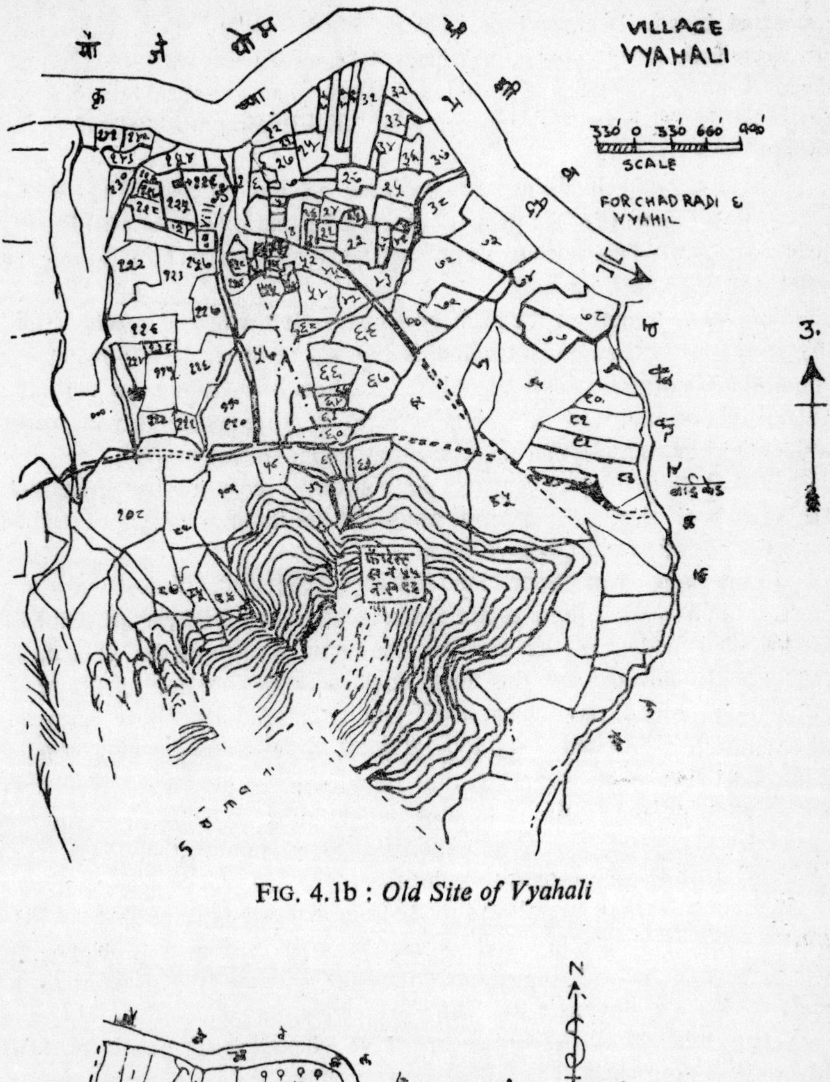

FIG. 4.1b : *Old Site of Vyahali*

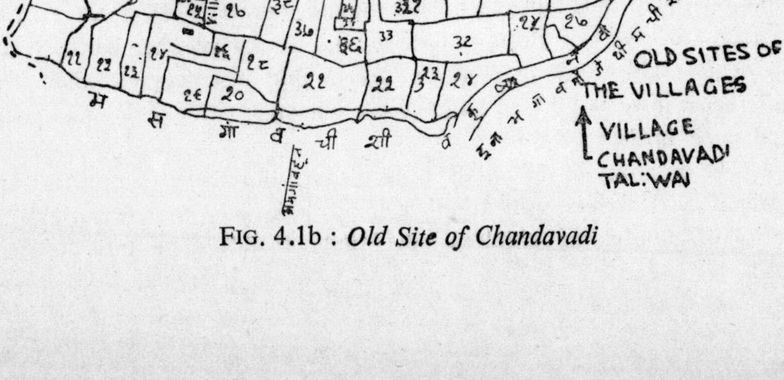

FIG. 4.1b : *Old Site of Chandavadi*

connected by small narrow lanes. Nearly 50 per cent of residential houses are electrified. Road connecting the village settlement with the Dhomeshwar Temple is made of cobbled stones. The annual fair of old settlement is celebrated in the name of God Bhairav on Phalgun Vadya Saptami and Ashtami.

The following facilities are available in old village: (Fig. 4.1a : (*i*) Post Office, (*ii*) One primary and one secondary school, (*iii*) Private dispensary, (*iv*) Two ration shops and one stationery shop; (*v*) one tea hotel, and (*vi*) One flour mill.

The total strength of the primary school is 190 with five teachers and the strength of the secondary school is 293, where 11 teachers (8 males +3 females) are working at present. All the teachers are residing at Wai and attend school daily by S.T. bus service, due to unavailebility of proper residential houses and other facilities, they prefer to stay at Wai.

In the village 70 per cent of population is engaged in agriculture. Main crop is rice. 20 per cent sample of the families have been studied from the unshifted old Dhom village.

In the category of "poor" land holdings (holding land below one acre), most of the backward families are landless labourers. Buddhas, Chambars are included under as the weakest economic group of the village, as they are also affected by the dam construction. The pure pressure of population which is unemployed and comes under above 16 years of age group is 13 per cent to the total population. Earning population is 41 per cent, supports the remaining 59 per cent dependents. Per capita income is Rs. 1.45 and landholding is below 0.34 ha. (3.4 Gundhas). They are covered very infertile, barren land of the village. They obtained their major incomes through the source of working as labourers in the village. They are always ready to accept any type of physical hard work. They are always in search of labour work inside and outside the village. Female work harder though their daily wages rate is less than the male worker, i.e., for male Rs. 5 and for female Rs. 3.50.

Government facilities and co-operative schemes reach hardly to them. Most are ill nourished. The school-going children are bare footed, having insufficient clothes to wear and no sufficient food to eat. Drinking hot water, without sufficient milk and jaggery is the morning tea of the labourer. Their huts tends to be ill kept and insanitary. However, they tend to accept their way of life as an illiterate old woman of Chambar caste (aged 60 years) said (looking towards nearby Buddhist people), "there are no differences in society based on castes or religions. Man is the one 'religion and caste' in the world".

The economy of the village is agro-based but there is no progressive agriculture. Careless utilization of land, accounts for lower yields. The ratio of school-going children per family is 0.8. The ratio of female dependents for the village is higher than the males, due to the women are not allowed to work anywhere except the household duty. The women of the family of poor labourers are tied with household work as well as labour work in the agriculture with lower rate of daily wages.

This is the present position of old village and it seems that no new development has occurred in it. The Rehabilitation Office at Satara, has informed them about their possible shifting in the near future at Joshi Vihir, because the old village comes under "Dam Protected area".

On 8th March, 1983, at the time of annual social gathering of the Narsingh High School, Dhom, the Headmaster, explained that they are also shifting their High School building at the upward northern side of the present old village; (Towards Abhepuri village area). However, it is possible to have hotel, industry at this old location due to beautiful environment, resulted by the construction of Dhom Dam.

OLD AND NEW SITES OF OTHER DAM AFFECTED VILLAGES (FIG. 4.2)

Asare

A biggest village in the valley was situated to the left bank of river Walki "Krishna Kamndul", (tributary of river Krishna). The village was located at a distance of 5 kms. westward from Dhom and 14 kms. westward from Wai Taluka. Population (1971) 1412 (612 M+800 F) and 1981=1054. Total area of the village was 204.42 ha. (494 acres). Jowar, rice was stapple food and only 61.91 ha. (153 acres) of land was under forest. The irrigated land for the village was only 10.50 ha. (26 acres). Wells was the only important source for irrigation. River and wells were the amenities for drinking water. There was a Post Office, and primary school, in the village. Panchagani, was the nearest market centre to the village. Important agricultural productions were rice and jowar (foodgrains). Mango, Jambhul, Phanas, were the other important forest resources. Mountain slopes of the northern part of the village supported some of the villagers for cattle grazing. Buffaloes, cows and sheep were important animals.

Village settlement was compact and most of the villagers were poor, backward and small landholders. Principal economic sources from agriculture was rice. There was no electric facility provided to the village. The village is now shifted at Kumathe, Taluka Koregaon, at a distance of about 68 kms. south-east from the old location (Fig. 4.2).

Asagaon

The village was located on the left bank of river Walki and connected by a road westward from Dhom at a distance of two metres, with an area of 1261 acres and population (1971) 1231 (557 M+674 F), and 1981=127. The total irrigated area was 160.63 ha. (387 acres) (irrigated by river and

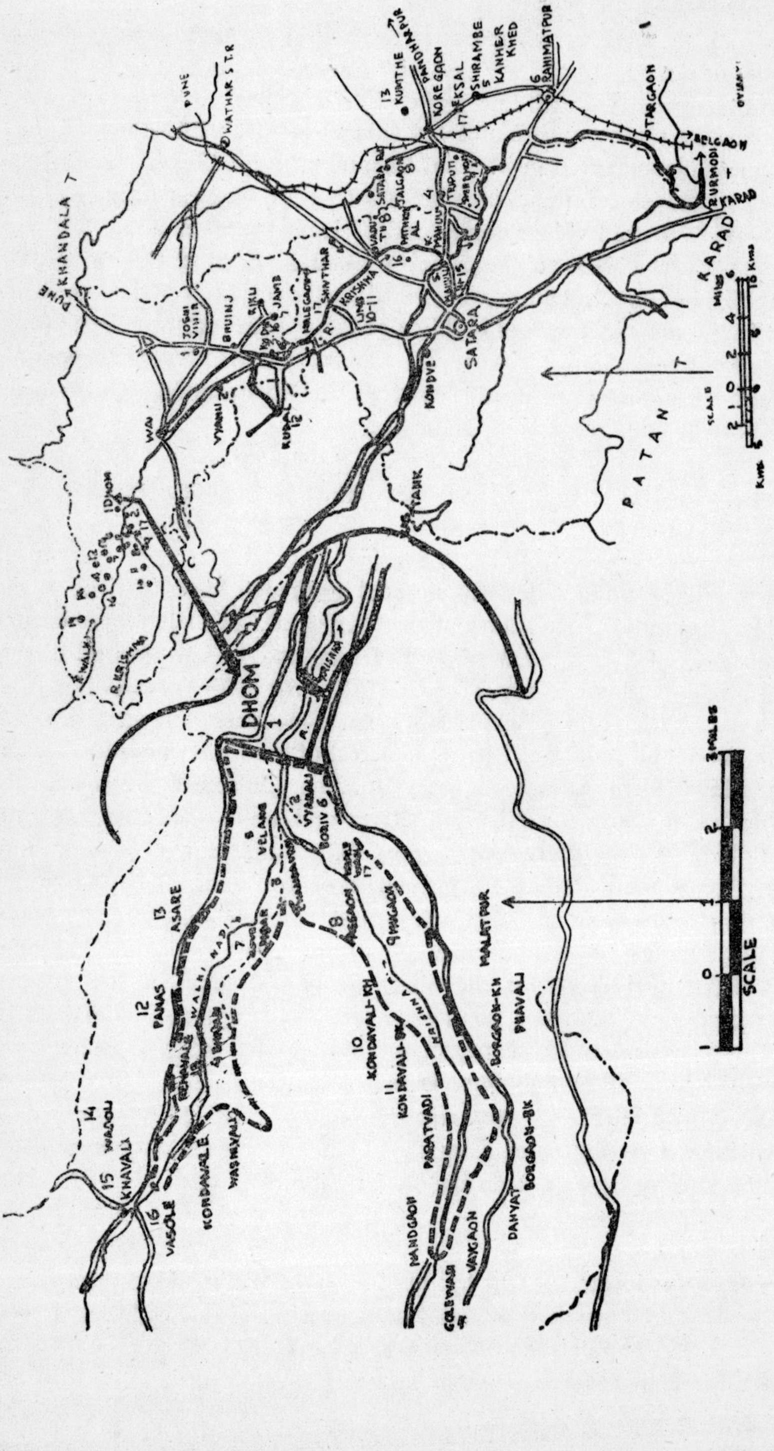

FIG. 4.2 : *Old and New Sites of Dam Affected Villages*

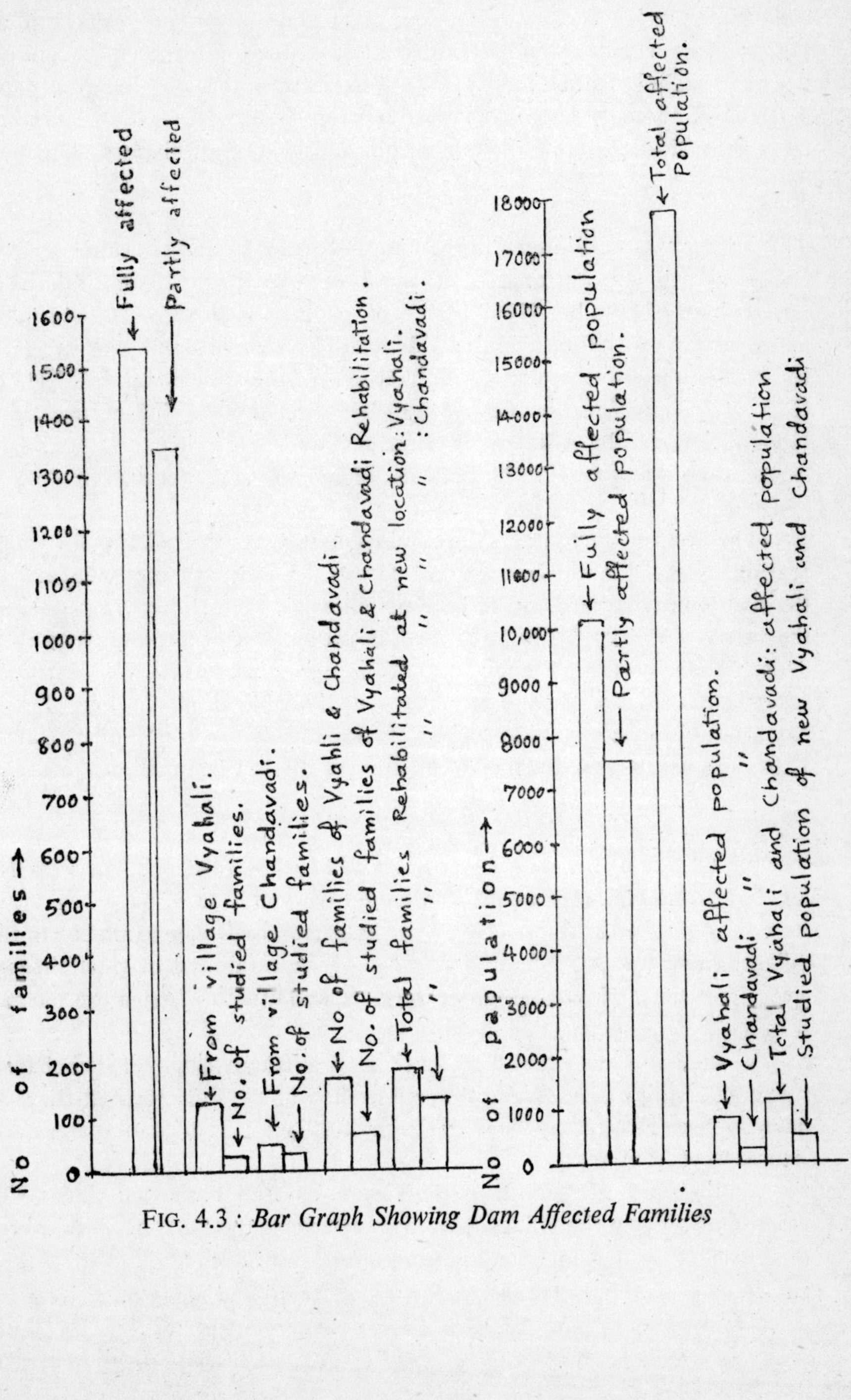

FIG. 4.3 : *Bar Graph Showing Dam Affected Families*

wells). Rice, ragi, and jowar were the food crops of the village. It was electrified village. There was one secondary and one primary school in the village. This village is shifted and resettled at the following two locations : (*i*) Near Vaduth Taluka Satara (South-east from the old location—at a distance of about 50 kms.), (*ii*) near Jalgaon Taluka Koregaon (South-east from the old location at a distance from old location of about 67 kms.)

Velang

This village was well-known in the Satara district for annual cattle market centre, which was arranged at the time of village annual fair. The village had 674.38 ha. (1669 acres) of geographical area, in which 196 acres, covered with forest. Only 17.37 ha. (33 acres) of land was irrigated by "well irrigation system". Population of the village was 1357 (1971) (596 Males and 761 Females). Major land was covered with cultivation of rice and jowar. The village was electrified; and Post Office, primary school were other facilities in the village. There was an ancient temple of "Maruti".

The northern mountain slopes were useful to the village for cattle rearing. River side was covered with forest. The soils of the village, from the river course towards north becomes infertile. The lower slopes along the valley were very fertile land, used for production of rice and vegetables. This village is shifted and resettled at three locations: (*i*) Near New Chandavadi: Bhuinj Sugar Factory at a distance of about 27 kms. South-east from old location; (*ii*) Near Kanherkhed; and (*iii*) Shirambe, Taluka Koregaon nearly at a distance of 60 kms. south-east from the old location.

Vyahali

The population of the village was 937 (1971) (463 Males and 474 Females) and had 374.69 ha. (887 acres) of area out of which 98.71 ha. (244 acres) were covered with forest. No well irrigated land was available in this village. There was 34.77 ha. (86 acres) of agricultural land irrigated through river irrigation. The water from the river was lifted for irrigation purpose through diesel engines (Fig. 4.1b).

Vyahali was located at the foothill of northern slope of Panchagani range on a spur extended and lowered in the valley to the north at the right bank to river Krishna and just opposite of village Dhom at a distance of 7 kms. west from Wai.

To the south of the village, there were covered mountain slopes with grass lands. From the village towards north up to the Krishna river course, the area was fertile, cultivated under rice crop, and was covered with mango, chinch, sag, jambhul, phanas, vad, pimpal, babul and many other trees.

The southern part of the village had infertile soil, under jowar crop which depended on monsoon rain only. There was a temple of Maruti in the village. There were 134 occupied residential houses and 135 households. Out of 135 households, 18 B.C. households were settled.

The original location of the village is now occupied by Dhom Dam-wall which has been constructed.

This is the first shifted village of Dhom Project. It is resettled at Kadegaon village area in Wai Tal. (Distance of about 13 kms. east from old location).

Chandavadi

The village was located above meeting place of Valki Nadi (tributary) and river Krishna, 13 kms. away (westward) from Wai and 4 kms. west from Dhom Dam. The settlement was located along the right bank of river Valki, and to the south-west of village Velang. The lower part of the village was very fertile agricultural land, cultivated under sugarcane, rice, and wheat crops. Population (1971) 314, (155 Males and 159 Females) with 81.34 ha. (201 acres) of total area occupied by the village. No forest land was available to the village. The irrigated land by river was 12.12 ha. (30 acres) and by wells it was 12.52 ha. (31 acres). There was a primary school and a temple of Shri Vitthal. The village was electrified. To the backward side of the village, there was a gentle slope of mountain spur which was backbone of the village supporting lower part for rice cultivation and the upward part for grazing land. Now, the slope is merged and dis-appeared within the stored water of Dhom Dam. The village area was bounded to the north by river Valki and to the south by river Krishna. The eastern part of the village was covered with forest. Forest products were an important source to the villagers. Settlement of the village was compact. There were forty-two occupied residential houses and number of households was fifty-one. There was no settlement of scheduled castes people. Un-irrigated total area was 35.99 ha. (89 acres). The area covered under culturable waste was 10.91 ha. (27 acres) and area was not available for cultivation was 9.69 ha. (24 acres). Velang was the market centre for this village (Fig. 4.1b).

New Chandavadi, is formed by mixed new settlement of the villages Velang and Chandavadi. Old Chandavadi is shifted from Walki Valley to south-east near Bhuinj sugar factory at a distance of about 32 kms. from old to new location. A very few households of old Chandavadi are shifted at Tadavale Tal. Koregaon (Fig. 4.2).

Boriv

Just above to the west of Vyahali, the village Boriv was located. The people of the village were very backward, and they had no contact with modern facilities. They were living subsistence type of life, depending on dry farming agricultural production on a very small patches of agricultural farms.

"Karavand" was also their subsidiary forest product. Rice and jowar, ragi were the main food crops.

The village had 188.27 ha. (466 acres) of land out of which only 34.37 ha. (85 acres) were irrigated by river and wells. Population was (1971) 567 (253 Males and 314 Females). No area was under forest for this village. Most of the economy of the village was depending upon the agricultural crops of rice, ragi and jowar food crops. There was a primary school, a temple of Maruti. It was electrified village and for it the nearest market centre was Wai. The village is shifted at a distance of about 65 kms. south-west from the old location at Rahimatpur area Tal. Koregaon.

Govedigar

This was a compact village, located to the right bank of river Valki, just opposite and southward of village Asare. The total population was (1971) 585 (284 Males and 301 Females), with total area of 269.51 ha. (666 acres). Total irrigated land of the village was 52.19 ha. (129 acres). There was a primary school, a temple of "Shri Maruti". The agricultural land was cultivated with rice, ragi and jowar food crops. The nearest market centre was Panchagani to the village. There was no extensive land under forest. Wells and river were the sources for irrigation. The village was not connected with other village by roads. It connected by simple footpaths, because it was located in the interior of the valley. Most of the villagers depended on sheep rearing. The northern fertile land of the village area was covered with patches of rice cultivation and the southern part was under jowar cultivation. The southern mountain slopes were covered with grasslands and used for sheep rearing purposes. The village is shifted from Walki Valley towards south-east and resettled at Kalangawadi village area Tal. Wai at a distance of about 52 kms. from the old location.

Mugaon

The village was located to the westward of Vyahali, near Nhalewadi village. It was located to the right bank of river Krishna and to the foothill of southern slope of Panchagani range. Along the river course area, there was rice cultivation and the southern part covered with grass lands. From this village, the fertile lower part is lost under the water reservoir of Dhom Dam.

The population of the village was (1971) 322 (168 M+154 F) and the village had an area of 213.26 ha. (527 acres), in which 51.38 ha. (127 acres) of land was irrigated by irrigation source of river, there was no forest area for the village. The village was located in a compact narrow valley and economy of the village was very poor. There was no electric facility in the village. There was "Chavaneshwar temple". The nearest market centre was Wai.

The village is shifted south-east, at a distance of about 63 kms. from old location at Shirdhon Tal. Koregaon. There were statues of Lord Ganakeshwar and Ganesh in the old village Mugaon which were shifted and kept carefully at the temple of "Kashivishweshwar" at Wai, (near the well known Ganesh temple of Wai).

Bhivadi

The village was located on a very infertile southern slope of Valki (Krishna Kamandul) Valley. The population of the village was (1971) 608 (264 M+ 340 F), and the total area covered by the village was 263.45 ha. (651 acres). Under forest 69.18 ha. (171 acres), and under irrigated area 49.76 ha. (123 acres) was covered. Rice, ragi and jowar were the crops produced from agricultural sector. Panchagani was the nearest market centre to the village. There was no post office and no electric facility in the village. There was Maruti temple and primary school. There were no proper roads connecting to the vlllage. Bushes of Jambhul, Karvand were scattered in the southern part of the village. The village is shifted at south-east near New Chandavadi Tal. Wai, at a distance of 32 kms. south-east from old location and 68 kms. away at Jamb Triputi village Tal. Koregaon.

Khavali

The village was located in the upper part of Valki Valley, on the left bank of Valki Nadi. The village had a very small agricultural land which was used to cultivate for only rice and ragi crops. To the northward of this village there was extensive mountainous area covered with grassland. The population of the village was (1971) 634 (287 M+347 F) and area covered by the village was 392.14 ha. (969 acres) out of which 62.31 ha. (154 acres) was under forest. The total irrigated land was 52.59 ha. (130 acres). There was a primary school and Post Office. Rice and ragi were important food crops. The village is shifted and resettled closed to the Bhuinj sugar factory, eastward from old location at a distance of about 32 kms.

Vasole

The village was located to the south of Khavali on the left bank of Valki Nadi. The poor village economy was depending on rice, ragi and jowar crops. Agricultural production was very less. There was a primary school and a temple of "Shri Maruti". There was Post Office and Wai was the nearest market centre to the village. The population of the village was

(1971) 1075 (478 M+597 F) and the area covered by the village was 2453 acres. The village is shifted at two locations—(i) At Pathakhal Tal. Satara (Distance from old location is 60 kms.), (*ii*) At Bhuinj sugar factory near New Chandavadi at a distance of 34 kms.

Kondavali Kh. and Bk.

Both the villages were located on the left bank of river Krishna. Population of Khurd was (1971) 284 (130 M+154 F) and of Bk. was (1971) 542 (245 M+297 F). Area occupied by the village was 233.10 ha. (581 acres) for Kh. and 369.48 ha. (913 acres for Bk.). There was 56.63 ha. (140 acres) under forest in Kondavali Kh. and 1280.45 ha. (3164 acres) in Kondavali Bk. The irrigated lands at Kh. were 42.49 ha. (105 acres) and at Bk. 26.69 ha. (66 acres). To both the villages, river was the only source of irrigation. To them Panchagani was nearest market centre and rice, ragi and jowar were the main crops. The villages were densely forested by mango, karvand, jambhul trees. To the south of the villages, there was very limited land cultivated under rice. There was a primary school to each of the above two villages. The villages are shifted at one location near village Limb Tal. Satara at a distance of about 37 kms. south-east from their old locations. Panas, Vadoli, Renewale, Dhavti and all the other villages were very small in size and population.

Old and New locations of Dhom Dam affected villages (Fig. 4.2)

Old location	*New location*
Fully affected villages	
1. Dhom	At Joshi Vihir-Ozarde, Tal. Wai.
2. Vyahali	Kadegaon, Tal. Wai.
3. Chandavadi	Near Khadaki Tal. Wai.
4. Bhivadi	(*i*) Near New Chandavadi Tal. Wai. (*ii*) Jamb Triputi, Tal. Koregaon.
5. Velang	(*i*) Kanher Khed and (*ii*) Shirambe Tal. Koregaon (*iii*) Khadki Tal. Wai.
6. Boriv	Rahimatpur Tal. Koregaon.
7. Gove Digar	Jamb and Kalangwadi Tal. Wai.
8. Asgaon	(*i*) Near Vaduth Tal. Satara and (*ii*) Jalgaon Tal. Koregaon.
9. Magaon	Shirdhon Tal. Koregaon.
10. Kondavali Kh.	Near Limb Tal. Satara.
11. Kondavali Bk.	Near Limb Tal. Satara.
12. Panas	Near Kudal Tal. Jaoli.

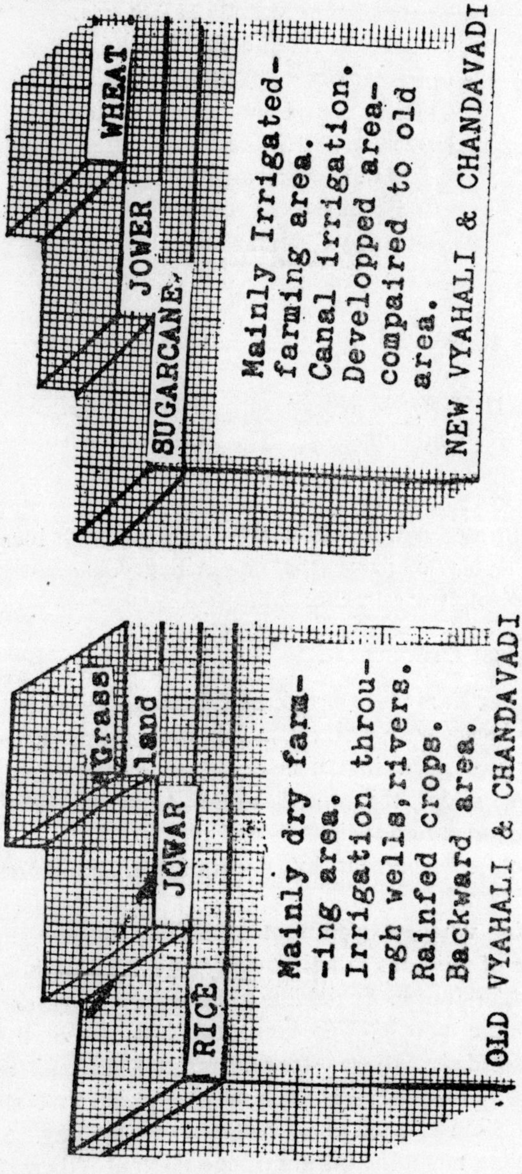

FIG. 4.4a : *Old and New Environment with Prime Crops*

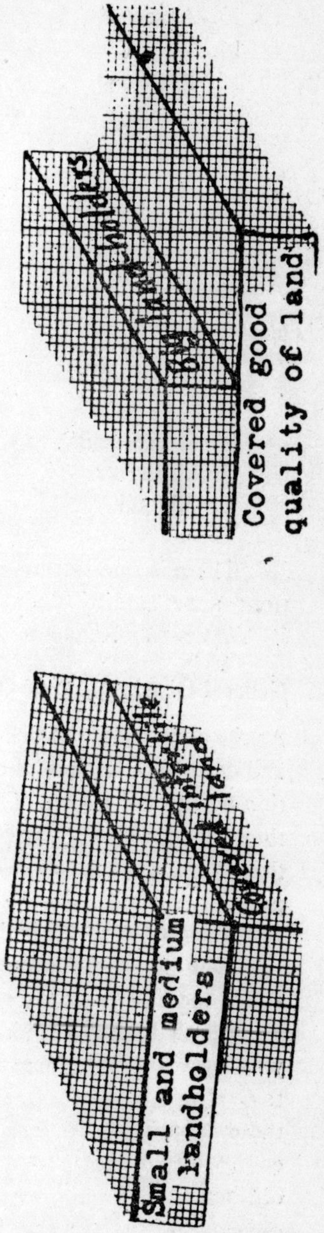

FIG. 4.4b : *Landholders and Their Occupied Land at Old and New Sites*

13. Asare	Kumthe Tal. Koregaon.
14. Vadoli	Near Sangam Mahuli Tal. Satara.
15. Khavali	(*i*) Sangam Mahuli Tal. Satara, and
	(*ii*) Near New Chandavadi Tal. Wai.
16. Vasole	(*i*) Near New Chandavadi (Khadki).
	(*ii*) Near Pathakhal Tal. Satara.
17. Nhalewadi	(*i*) Malgaon-Shivathar, Tal. Koregaon.
	(*ii*) Eksal (Near College) Tal. Koregaon.
18. Renawale	Near new Chandavadi Tal. Wai.

Unshifted Villages

A. Jambhali	B. Washivale	C. Kondavale
D. Balkwadi	E. Nandagane	F. Partawadi
G. Borgaon Bk.	H. Borgaon Kh.	I. Dhavali
J. Malatpur	K. Chikhali	L. Dasawadi
N. Akoshi		

There is one settlement formed by the following villages at new location: near Bhuinj Sugar Factory close to the new village Chandavadi (*i*) Vasole, (*ii*) Renavale, (*iii*) Khavali, (*iv*) Bhivadi.

General Observations of Original Environmentl

As the construction work of the dam was started the villages appeared pale and deserted. Some of the villagers were working as labourers for construction work of the dam. They knew that in the near future everything in their villages would disappear under the stored water. And well before that, they had to shift somewhere on their own and they would be compulsorily shifted out of this old environment by government. The houses and huts began to crumble down.

Due to uncertainty, the villagers were not able to participate and enjoy themselves at the time of religious functions and festivals. Everybody was puzzled with an uncertain future. Social life became inert and worry writ large on the faces of the people. The landscape lost its vitality, the temples, the once crowded village streets, mango gardens, all these appeared to lose their functions. It was also noticed that the energy of working persons of dam affected area decreased day by day. An old person at Velang was trying to soothe his mind; he said that "there is only one world; the sky is the same, the land is also the same, which belongs to the earth. We have to change our location from here to other place".

"हे विश्वची माझे घर". He said, further looking towards the dam site, the water of the dam is going to attack and capture all his environment, for which he had spent his life. He cultivated mango trees and for the last 20 years, he was getting income from the trees. Under a very sad mood, he explained that he was anxious about how and where his next generation would adjust the life problem of settling at the new location. He was worried about the future of his family.

By 22nd August, 1982, the dam was full with water. On this day, from village Umbarwadi (which was included in a part of old village Velang), it was a view of over looking the top branches of a tree near the temple of old Chandavadi, above the surface of the dam water. Umbarwadi, a village, which is still settled at the original location, and at 15′ distance to south from the residential buildings of Umberwadi there reaches the maximum water level of Dhom Lake. Some farmers have shifted to protected upward mountain slopes and started to construct new buildings for them. Dagadu, Baban and Waman Jadhav, the landholder of Umbarwadi said, "in rainy season when the water level of Dhom Dam reaches near to their residential sheds, the snakes from nearby area gathered around their settlements to take shelter from water of the dam and from the rain. The soil of the land being wet, swelled and puffed upward, resulted many cracks and faults to the walls of their houses". No residential plots are acquired by the government for the project from Umbarwadi. No agricultural land is given to any of them. The compensation of Rs. 2000 per acre is given to small landholders. Only one family of this area have obtained agricultural land at Khanapur village, at a distance of 8 kms. east from Wai.

All the affected villages, were economically backward and poor. The villages were important for rice production. There were no proper roads in the region. Bullock-cart was the only communication facility available in some villages, but to the western part of the valley, the villages of Vadoli, Khavali, Washivali, Vasole, Kondavale, Nandgani, Partawadi, Golewadi, Vayagaon, Dahyat were connected by only footpath or cart tracks to each other.

Before electric facility, most of the villagers used kerosene lamp. Grampanchayat provided protected kersosene lamps on a pole with glass covered boxes on every corner.

It was very interesting to see the social and other functions through petromax at night.

Generally rich person (i.e. Patil of the village) used a horse for travel.

"Dalge" (roundish carrier made of bamboo sticks) was very helpful to carry the sick persons from the village up to Wai. Still for the present, villages like Baleghar, Mandhardeo, (located at the top of northern mountain range of Wai taluka) utilise "Dalge" for carrying sick persons up to the Wai Hospital centre, due to lack of communication facilities, as they are located in the mountainous area.

In old times, the marriage ceremony was celebrated for five days without any difficulties of food supply and a procession of "Varhad" was taken through many bullock carts with five to six persons in each. "Varhad" is used for the people on both sides of the marriage to attend the marriage function.

All the affected villagers were living in a united manner. The society was rigidly grouped on caste systems. Their basic needs were very limited and they felt comfort with their little needs. Radio was the only entertainment.

The area was very important due to availability of vast extensive grass lands within the valley for dairy industry. Collecting gum, fruits, honey and other forest products was also supporting occupation to the weaker section of the community. Oil engines were used everywhere for lifting up water from river and wells for irrigation.

One of the person of village Lohare remembered his village, before the provision of electric facility. It was drowned in dark and calm atmosphere at night. He studied in a hut under the kerosene lamp. To travel from one village to the other at night was a strange experience. The poor villagers were able to maintain their basic needs without any difficulty during the period of 1950-60. It might be possible due to very limited needs of the families.

Bullocks were the important source of energy to the farmers in farming operations.

The villagers had to walk from one place to another specially to westward of Dhom to get the basic needs and daily requirements for the family and their main transport facility was bullock cart. Many villages were connected by footpaths. No proper roads were constructed within the valley. In rainy season, it was very difficult to pass through these unbearable routes. The region of the valley was not productive for agriculture. Rice, ragi, were the chief crops of the region and "kumari" cultivation was common and that was also mostly concentrated to the lower slopes of the valley.

The sizes of the farm have been very small till now. All the villages were with compact type of settlement pattern. Being mountainous and economically weak, such area could not get medical facilities. Snake bite was common in the region. Natural calamities were solved by traditional ways. Snakes bites were cured by keeping the victim in the village temple. They believed God to be the curer of the victim. They are unaware of two things—(*i*) Psychological and mental power, (*ii*) Non-poisonous species of snakes.

They were under tension of God and Power. It was He, who kept them out of bad habits.

The Maratha is the only chief farming community of the area. Small and medium landholders with small patches of landholdings have covered the major part of dry farming area.

The settlement were clustering compact and traditional affinity was observed within the village.

Rice was the golden crop to the villagers in the Upper Krishna Valley of westward to Dhom Dam. It was a pleasant atmosphere that harvesting of rice and jowar is made by all the family member of the farmers by singing group songs called locally as "Irgik".

It was socially and culturally very important to arrange Bhajans, Grantha Samapti, Shahiri Povadas, Gondhals and such other programmes, which were useful to them for being more religious and behaviour of rural society is controlled by such religious traditions.

It is the present position after 1965, that with increasing family size, land fragmentation becomes common and small size of landholding are now affected badly with its productivity which is being diminished fastly. Land has been the main reason of most of the quarrels in such families. Family quarrels becomes common in the rural area which takes place due to land property only. Courts in cities have been crowded with such land cases. Such quarrels help to make economic condition worse and disturb peace of mind in the village.

All the affected villages were located far away from Wai taluka place and it was very difficult and expensive for the villagers to have immediate contact with doctors for medical help and for postal and other administrative work. These people could not develop relationship with eastern developed part of Wai Taluka.

Here, in Maharashtra, for Dhom project, the affected villages, have shifted without giving too much trouble to the government, but the experience of "Sri Sailam Project", in the State of Andhra Pradesh is very strange. The government staff of revenue, irrigation departments, along with hired labourers, with "Police Bandobast", has taken the step of demolishing the houses by removing doors, roof tops, walls of farmers.

New Environment

All the dam affected villages are resettled within the command area of the project. This part is quite developed part compared with its original environment. It has a capacity to provide through nearby villages, agricultural labour work to the new comers. Only the fact is that they have obtained improper and infertile land. To adjust within the new cropping pattern, to cultivate the land which is far away from their residential part and to obtain basic needs for the new settlement, are the challenges before them. They have to develop the new area by all means even in the fields of social, cultural, economic life of the community. They have to start to cultivate the natural beauty of their area by intensive agricultural methods and by cultivating trees and other suitable resources within the new environment. We should provide them land, resources and services also.

Shifting motivation was purely economic and due to shifting of settlements, environmental relationship of community life was disturbed.

It is remarkable that dam affected villages resettled at Koregaon Taluka are still (May 1983) in a very difficult position, fighting against the problem of supply of drinking water. The water of Dhom canal still cannot reach to support them due to inadequacy of water within it and due to incomplete construction work of canals.

5

Procedure in Land Acquisition and Rehabilitation

IN THIS chapter, we outline the government procedure in land acquisition for the purposes of project and rehabilitation.

Procedure

For an irrigation project, land is acquired in the following manner :

(*i*) Acquisition of agricultural and settlement area which comes under dam submergence.

(*ii*) Acquisition of project benefited (command) area for distribution of agricultural land and house plots to the affected persons shifting of affected families by providing them within suitable agricultural land and residential plots on the command areas.

(*iii*) acquisition for resettling the villages as a whole: selection of proper locations for settlement area in the benefited part of the irrigation project.

Land Acquisition in the Catchment Area of the Dam Reservoir

The land and houses are acquired under the Land Acquisition Act (1860 LAA), Sections 4, 6, and 9. Owner is required to inform and state the amount and particulars of his claims for compensation (as per Section 9 of the Act). Within a specific time, all the persons concern losing their lands under submergence of a dam have to contact and inform the Land Acquisition Officer (LAO) about fixation of value of his property (i.e., houses, trees, land, wells, etc.). If a person fails to do so, he shall not thereafter be entitled to claim more than what the LAO awards. If the owner or landholder, wishes to make any complaint against the land acquisition or the LAO, he has to file a suit in the Civil Court under the L.A. Act, Section No. 18. The Civil Court is the final authority to take the decisions on any rehabilitation or Land acquisition problems.

Temple lands are acquired under the Religious Endowment Act. The persons concerned with temple and religious and land property of any "Deosthan Trust",[1] have to contact within 60 days of the declaration date, to the L.A.O. for obtaining compensation of Deosthan Property. If they fail to do so, the claim for property and amount becomes invalid and the responsibility lies with the Government to construct a building for temple in the rehabilitated village. There is also provision in the Act to pay a fixed amount to the new settled village as alternative for constructing such temples.

Land Acquisition in the Command Area

Generally the affected villagers have to be rehabilitated within the command area of the project. The scale of acquiring land by government is given in Table 5.1.

The land is acquired from the landholders of command area, (above landholdings of 3.23 ha. or 8 acres). The small holder (i.e. 3.23 ha.) is protected. The bigger holder has a right to surrender a piece of his choice of land (and usually this is the least fertile part). As per rules, complaints by landlords about land acquisition have to be preferred to the Collector of the district who is final authority in this regard.

TABLE 5.1 : **Scale of Land Acquisition from Command Area**

Landholdings in the Command Area	Land Acquisition : Scale or Ratio
1. Up to 3.23 ha. (8 acres)	Nil
2. 3.23 to 4.84 ha. (8 to 12 acres)	Out of more than 3.23 ha. (8 acres) land should be acquired not more than 0.80 (2 acres).
3. 4.84 to 6.44 ha. (12 to 16 acres)	Out of more than 4.04 ha. (10 acres) land should be acquired not more than 1.2 ha. (3 acres)
4. 6.44 to 8.16 ha. (16 to 20 acres)	Out of more than 6.44 ha. (16 acres) land should be acquired not more than 1.61 ha. (4 acres)
5. 8.16 to 9.77 ha. (20 to 24 acres)	Out of more than 8.16 ha. (20 acres), land should be acquired not more than 2.02 ha. (5 acres)
6. More than 9.77 ha. (24 acres)	Out of more than 7.68 ha. (19 acres), all the land should be acquired.

REHABILITATION OF DAM AFFECTED PERSONS

For Completely Affected Villages

If more than 75 per cent of the settlement and other constructed buildings of the village are acquired, or the complete village settlement area is acquired with undertaking written of the all affected villagers.

In a village where more than 75 per cent of agricultural land excluding settlement area is acquired, if the land is cultivated after 15 December in the year, and if the total calculated area comes to 50 per cent of the total cultivated area, the village is called as "*fully affected village*". Such fully affected village is shifted and rehabilitated as government responsibility.

For Partly Affected Villages

If some part of agricultural and settlement area is acquired for an irrigation project, which is less than that of the above mentioned ratio, the village is called as "*partly affected village*". Such a village cannot claim as a right and to obtain government grant for resettlement purposes. Partly affected villages are not shifted and resettled at government cost. They are not provided settlement plots within the new site. These partly affected villages lose mostly their fertile land under submergence. However, if any such partly affected village demands rehabilitation facility and if their present land is mostly barren and infertile, there is a provision with the government that after examining their difficulties, the government may take the responsibility of rehabilitation of partly affected village. It is always observed that partly affected villagers fail to explain their difficulties and problems with proper reasons for obtaining rehabilitation facilities.

The government has accepted to provide rehabilitation of fully affected persons of partly affected villages.

Rehabilitation for Landless Labourers and Village Artisans of Project Affected Villages

Labourers and village artisans, though they have no landed property as such within the affected villages, and those who are landless poor workers, are included under affected persons and given facilities of rehabilitation by providing them free of charge residential plot and one acre of agricultural land to each family. But, grocers traders, shopkeepers, and merchants and traders or persons of same category, and those who are economically in sound position are not called as dam affected persons.

Definitions of Dam Affected Persons

(*i*) Those who have lost their agricultural land and residential buildings for the purpose of project, and

TABLE 5.2 : **Scale of Distribution for Land from Command Area with Acquired Land of Catchment Area, to the Affected Persons**

Total land acquired for the project from catchment area	Land distribution to the affected persons: from the command area
1. Up to 0.80 ha. (2 acres)	Up to 0.80 ha. (2 acres)
2. 2 to 5 acres (0.80 to 2.02 ha.)	Up to 1.21 ha. (3 acres)
3. More than 2.02 ha. (5 acres)	Up to 2.02 ha. (5 acres)

(*ii*) Those who have lost their above property of land and buildings under reservoir, and

(*iii*) Those who have lost their land due to land acquirement by government for residential buildings under project roads, canals, colonies, dam-wall are called the project affected persons. In the benefited area (*a*) those who have lost their land under project, and only holding below one hectare (2.5 acres) and (*b*) those whose agricultural farms were lost for canals and the remaining land becomes unsuitable for cultivation are called as "dam affected persons."

Land Distribution

While distributing the land to the affected families, the number of family members are considered as a base for distribution on the following scale:

If there are more than eight family members, for additional each of three members 0.40 ha. (one acre) extra agricultural land should be given up to maximum limit of 2.83 ha. (7 acres).

Now, government has improved the scale of distributing land to the affected families and have decided that in a family; with more than five family members, for each additional three members 0.40 ha. (1 acre) of extra agricultural land should be given up to a maximum limit of 2.83 ha. (7 acres). This scale came into effect from the 4th October, 1980. According to the L.A. Act, Section No. 4, the number of the family members of the landholders of the command area is furnished as per the names registered in the Ration Card, for land acquisition purpose.

After completion of the work of distributing compensation in the form of land to the affected persons from the command area, available extra and additional land is distributed to landless labourers and to the village artisans as per ratio of 0.04 ha. (1 acre) land to each family.

While distributing land to the landholders, "the name of the landholder" (the head of the family) and his dependents are considered as "one family" as per land record in the village abstract. The amount should be kept with the government as a deposit from the following alternatives: (*i*) Price of land from the command area or (*ii*) 75 per cent of price of acquired land by the government from catchment area by any affected family. Without this no family qualified to obtain alternative compensation of land for rehabilitation. Of course, from the total amount of compensation only 25 per cent of amount is returned to the affected land holders and the remaining amount is utilised for giving him available and acquired land from the command area. Of course no problem arises in respect of collecting amount from the landless affected persons of old locations.

According to the scheme of distribution of land to the affected persons, the land is given on the condition that the land is not subdivided into members of the family. Without prior permission of the District Collector, such land is not allowed to be sold or divided or mortgaged or transferred to anybody. Termination of such agreement is liable for legal action.

Method of Land Acquisition of Command Area

Special Land Acquisition Officer, as per Section 4(1), gives notice to the old landholder of command area for acquiring his land. Then he gives another notice under Section 6(4) asking for any objection from the original landholder within a specified time. If no objection is raised by him, the land is declared and published in the Gazette. Then according to section 9(4) a public notice is given again if there is any objection of acquiring the land. If no objection is raised by anyone, a final notice is given to the owner: landholder under Section 12(4), and the amount of compensation is given to him as per government rules. The land is cultivated for some years by the original landholder, though the land is in the name of the government. This acquired land with its record is given in possession of Rehabilitation Department. After 1978, under Resolution No. RPK/1078-555/R 11/dated 15-11-1978, it becomes very difficult to obtain possession of agricultural land to the dam affected persons, from command area, because original landholder goes to the Court of District Collector or to the Commissioner and finally to the Government of Maharashtra for unwilling to give possession of his land to the affected persons and provides some documents being a divided family as small landholders. In this process two or three years are spent. This makes process of rehabilitation work lengthy and in this way possession of agricultural land from command area becomes more complicated.

Distribution of Residential Plots

Partly and fully affected villagers are given alternative new locations for settlement purposes. In general infertile land is selected for settlement purposes, and fertile land for agricultural purposes.

For the selected new location, the number of plots with the number of affected families, are shown by a layout map, prepared for suitable

TABLE 5.3 : **Distribution of Land for Residential Plots (up to 11-1-1978)**

Number of affected family members	Minimum area	Maximum area
A family of land holder with 5 or less than 5 members	4,000 sq. feet up to five members and in addition for per 3 persons +1,000 sq. feet, but not more than up to the fixed maximum limit	6,000 sq. feet
A family of affected non-landholders having five members	2,000 sq. feet, and in addition to each 3 persons +500 sq. feet up to the maximum limit.	3,000 sq. feet

TABLE 5.4 : **Revised Rate of Distribution of Land for Residential Plots (from 12-1-1978 and onwards)**

Number of affected family members	Distribution of area for residential plots	
	Minimum area	Maximum area
Affected landholders having less than 5 or 5 family members	4,000 sq. feet up to 5 members and in addition for each 3 persons +2,000 sq. feet up to maximum limit	8,000 sq. feet
A family of affected non-landholders having five members	2,000 sq. feet, and in addition to 1,000 sq. feet to each 3 persons up to Maximum limit.	4,000 sq. feet

distribution of plots. Land for residential plots are distributed as per Table 5.3

The above rule was revised and changed by the government on 12.1.1978. From this date the following scale is fixed for distribution of plots to the affected families as given in Table 5.4.

As per above scale, plots are distributed with reference to the following considerations :

(*i*) The affected person (landholder) should have a residential house in the old village.

(*ii*) Members of a joint family, are not called separate landholders.

TABLE 5.5 : **Distribution of Loan in Stages for Construction of a House**

Step of house construction	Instalment No.	Amount obtainable in Rs.	For non-landholders amount obtainable in Rs.
Before starting the construction	First	800	400
Construction completed up to plinth	Second	2,000	500
Construction completed up to Roof of the house	Third	2,200	600

(*iii*) If a landholder, is regarded as an independent separate landholder by the court, he is entitled to hold a separate plot from the government in the new location.

Facilities Provided in the New Settlement of Rehabilitated Villagers

For each affected family, a loan of Rs. 2,000 is sanctioned for construction of a house. For each affected family on non-landholder, a loan of Rs. 500 is sanctioned for the same purpose. Amount of this loan is given in three instalments. The amount of loan has been revised from 1.4.1980, for each affected family. Rs. 5,000 is sanctioned and for the family of affected non-landholder, Rs. 1,500 is sanctioned as a loan and help for construction of house. The amount of loan is given in the following ways, in three instalments.

Those who have been granted but not given the amount, are liable to get the amount in the revised scale sanctioned on 1-4-1980. The amount is given in instalment by the B.D.O. (Block Development Officer). But the owner of the land has to give a registered mortgage bond. The work of building construction completed up to plinth or roof, has to be certified by the overseer of the B.D.O. for obtaining the next instalment of the grant.

Facilities Provided to the Rehabilitated Villages

Land and residential plots are allowed by the District Collector and the Asstt. Director of Rehabilitation. All the civil amenities are provided by the Chief Executive Officer of Z.P. and by the Asstt. Director of Rehabilitation. The amount required for such facilities is sanctioned to the Z.P. by the Project. So, the responsibility lies with the Z.P. to provide all the required facilities to the affected village. Deputy Engineers are appointed under the Chief Executive Officer of Z.P. for the work of supervision and implementation.

TABLE 5.6 : **Number of Wells Provided to the New Settlements as Per Population of the Village**

Population of the new location rehabilitated village	Number of wells provided to the new settlements
Up to 1,000	1 tube well and 1 well with 15′ diameter
1,000 to 1,500	1 tube well and 2 wells with 15′ diameter of each
1,500 to 2,000	1 tube well and 3 wells with 15′ diameter of each
2,000 to 3,000	2 tube wells and 3 wells of 15′ diameter of each
Above 3,000	2 tube wells and 4 wells (15′ diameter)

Major facilities to be provided to the new village are :

(*i*) Water for drinking purposes,
(*ii*) Agricultural land within the distance of 8 kms. from the new village location,
(*iii*) Roads within the new settlement and roads connecting nearby villages.

Drinking Water Facility

To every rehabilitated village, government provides drinking water facilities on the scale as given in Table 5.6.

If water provided by wells is found insufficient, water facility through pipe line water scheme is provided to the village at government cost. Till then, drinking water is provided by tankers to the village.

School Building

Considering the strength of school-going children in the new settlement, school building is constructed with cost.

Play Ground

For primary school, one acre; and for secondary school, two acres of land including play-ground is provided in the new location.

Grampanchayat Building

Grampanchayat building or recreational social work building is provided by the government.

Roads

Roads are constructed within the new settlement.

Common Cattle Shed

For each affected village one acre of land is provided for common cattle shed and grazing land.

TABLE 5.7 : **Scale for Providing Land for "Khalwadi"**

No. of persons rehabilitated in the new village	Land provided for Khalwadi
Up to 200	0.20 ha. (20 Gunthas)
201 to 400	0.40 ha. (1 acre)
401 to 600	0.60 ha. (1 acre and 20 Gunthas)
601 to 800	0.80 ha. (2 acres)
801 to 1,000	1 ha. (2 acres and 20 gunthas)
Above 1,000	1.21 ha. (3 acres)

Provision of S.T. facility and S.T. Pickup shed, (if this request is made to the State Transport Corporation).

Cremation Ground

Cremation ground or burial land. There is a provision of half acre of land provided to each religion for cremation ground, i.e., Hindus, Muslims, Christians etc.

Provision of an enclosed place outside a village, containing the thrashing floors of several cultivators is made; it is known in Marathi as Khalavadi (खलवाडी). If in the original location, place for Khalwadi is available then in the new location land for Khalwadi is provided on the scale as given in Table 5.7.

Clean and Open Gutters

Gutters are provided within the new settlement.

Grazing Ground

Government can provide such ground to the new settlement, if available in the area. But it is not binding over the government to provide and acquire land for grazing purposes.

Common Lavatory and Privy Facility

Nearly half acre of land in the new location at suitable place is reserved and provided to each village of new locations for Common Sanitory Blocks.

Supply of Electricity

It is the responsibility of the project authority to provide at their cost electricity to the new village.

Facilities Provided to the Affected Villagers or Displaced Persons

Compensation is given for acquiring old land and other property (i.e. trees, wells, etc.) which comes under dam submergence.

Free transporting facility is provided from old to new locations to transport doors, windows, house materials and wood. Cutting of trees would be done with the permission of revenue or forest officer.

Concession of Stamp Duty

No stamp duty is required for taking possession of lands, according to legal process of rehabilitation.

Cement, tin sheets, and such materials are provided to the affected persons. To each family 10 cement bags are provided in the new location.

Concession for giving Employment in Government Services

Project affected person and his dependents have been given preference for government employment in Class III and IV services (in the Irrigation Department, Rehabilitation Dept., Revenue Dept., etc.). For affected persons, no employment exchange registration is required. Age limit for Govt. Services is also relaxable up to three years. They are confirmed immediately in services. Affected and displaced persons are provided certificates as "Dam affected persons", for employment purposes. A certificate is provided to one family and affected persons are called with his dependents, included in it, his wife, son, daughter, unmarried sister, brother, father, mother or husband. Such certificates are obtained by applying in the prescribed form to the L.A.O. The names are mentioned by Rehabilitation Office in various departments of the district for giving them preference in services. Such certificates are not assurances for services. It is the responsibility of individuals to obtain services.

Five per cent of seats are reserved for the dam affected persons in the government services. Interest of 4 per cent is given to the affected person, who has deposited with the government the prescribed amount for obtaining agricultural land within the new area. If the price of the land in new location is more than that of the price of the land of old village which is acquired by the government for project, then this excessive amount of new land is collected from the affected person in three annual instalments.

Loan by mortgaging land for agricultural development can be taken by the resettled farmers, from L.D.C. or equivalent banks.

Duties of Project Affected Persons

It is the duty of every landholder, to ensure an up-to-date record in the new rehabilitated village abstract whether noted by the Talathi of the village. From the rehabilitation office, intimation is already given to the related village Talathi to record the name of the person as a landholder with his new land and house plot property in the rehabilitated village. The affected landholders are provided free of charge a copy of G.A.N. 7/12 of village abstract.

TABLE 5.8 : **Present Administrative Pattern**

Responsibility and duties of the Govt. offices in relation to land acquisition and Rehabilitation work	Concerned officer to whom the applications of request about any difficulties should be made by the dam affected persons
1. To acquire the land of submergence and command area and to give compensation of the land: (*i*) Land acquisition from command area to give compensation of alternative land to the project affected persons (*ii*) Land acquisition of submergence area from affected persons	Concerned Land Acquisition Officer
2. Land acquisition for new village settlement locations and to give compensation for acquired land	Concerned Land acquisition Officer
3. Land Distribution to the project affected persons	District Rehabilitation Officer
4. To select new location for a village and to distribute plots to the affected villagers	District Rehabilitation Officer
5. Shifting of affected villagers	District Rehabilitation Officer
6. To locate alternative land and to survey it	District Rehabilitation Officer
7. To provide facilities to the village at new locations	District Rehabilitation Officer
8. To solve the problems of acquired land and to remove obstacles of the land holders for possessing the new land	Concerned Mamalatdar
9. To provide drinking water through pipe line water schemes	Executive Engineer
10. To provide water, road, school and other facilities in the new village area	The Chief Executive Officer Zilla Parishad
11. To supply electricity.	Maharashtra State Electricity Board
12. Supply of cement, tin-sheets to the affected persons	Executive Engineer of the Project
13. Distribution of loan for construction of residential shelter within the plots	Concerned Block Development Officer
14. Pass Book of land record and village extract of VII-XII (for alternative land).	Village Talathi.

Transfer of Land by the Old Landholder to New Landholder

It is generally found to bring obstructions while taking possession of land given to the affected persons by original landholder of command area. The original landholder should be present on his agricultural land to show the signs and marking of the boundary. If he remains absent at the time of surveying his land, the land is given in his absence by the L.A.O. to the affected landholder. If any objection is raised for giving possession to the affected person, then he has to make a complaint against the original landholder under the Mamalatdar court act under Section 5. Under this Section, he can request the Mamalatdar or Talathi for removing the obstacle of getting possession of acquired land for rehabilitation within six months.

Land Acquisition for Canal Purposes

For building the canal routes, land in the command area is acquired by the government. Under project work, on or after 1.1.1980, if the land is acquired as per Land Acquisition Act 4, and if such landholder who has lost his land under canal construction and if only less than one hectare of land remained with him then such landholder is called as "affected person" and alternative land is given to him by the government. He is also eligible to get "a certificate for government services as Dam affected person".

District Rehabilitation Officer is the chief co-ordinator, who co-ordinates all the concerned irrigation, land acquisition and rehabilitation offices.

Present administrative pattern of the land acquisition and rehabilitation is given in Table 5.8. The government has tried to solve the rehabilitation problem by shouldering the work and responsibility through different agencies that are financed by the Irrigation Project.

REFERENCE

1. Revenue or land assigned for the maintenance of any religious establishment—revenue which is dedicated to support temple (sacred place) is known sa "Deosthan Trust", or "Deosthan Inam".

6

Detailed Study of the Village Vyahali

Old Vyahali

THE OLD location of Vyahali was in the south-west of Dhom village (Fig. 4.2), and now it is covered by the water of the dam and situated at the southward constructed wall of the dam. Population of old village was 937 (463 males, and 473 females) in which scheduled castes population was 39 (in 1971). Now south of the dam wall, there are 15 houses which remained in their old site due to their upward location along the slopes of Pasarni Ghat. The village area was 358.94 ha. (887 acres) with a primary school and electricity facility. Main crops are, rice, ragi, and jowar. The area covered by forest was 98.71 ha. (244 acres). No area was under well irrigation. Only 34.77 ha. (86 acres) were irrigated by Krishna river and area of 171 ha. (425 acres) was under unirrigated land. There were Nath, Shankar, Vitthal, and Maruti temples and Wai was the nearest market centre for the village. At the old location, the village had 134 occupied residential houses with 135 households. Due to construction of the dam wall, the village was shifted as an emergency work very early, and this is the first village from all the affected villages which immediately came to the new location at Kadegaon in the year 1971.

New Vyahali

One hundred eight households with 412 population (213 males+199 females) (included with 18 B.C. families) are resettled at the new location which is west of Asale Bhavani Temple hill (Fig. 6.1 and 6.2). It lies in the southern area of Kadegaon village at a distance of 6 kms. east of Wai-Satara Road (on 73°56′E Long. and 17°54′N. Lat) (Fig. 6.3 and 6.5). The residents were shifted with their house materials by trucks provided with free of charge by

Irrigation department. The statues of Nath, Shankar, Vithal and Maruti were kept in one temple constructed with the help of the government.

The annual fair of the village is celebrated, (as it was celebrated at the previous location on the same date) on Kartik Vadya Ashtmi. The amount of Rs. 2000 is contributed for this celebration. Tamasha (Loknatya), wrestling programme and God Bhairav Palkhi procession are the main entertainment programmes of this annual Jatra.

Educational Facilities

At the new location, there is a primary school. A new high school is opened in 1983, within the building of Gram Panchayat Office.

For College Education the students have to go to Taluka place, i.e., Wai.

Agricultural Land

The new Vyahali, is bifurcated by obtaining agricultural lands of seven different villages (Table 6.1).

The total area covered under new Gaothan is 8.78 ha. (21.30 acres). (Fig. 6.3). This comes under infertile lower slopes to the south of Vairatgad range. The slope are covered with scattered bushes and grass-lands. The right bank canal which is life for this area, changing the original dry landscape into irrigated agricultural land along both sides of Wai-Satara road. The area which was originally under jowar, tur and other dry farming crops is now turned under sugarcane and vegetable gardening.

New Settlement (Fig. 6.5)

Out of 108 houses, forty-nine have electricity. There is settlement of B.C. (Buddhist) people at a distance of two hundred feet to the south of the

TABLE 6.1 : **Agricultural Land Obtained to New Vyahali from Different Villages (to Selected Sample Families) (Fig. 6.7 & 6.9)**

Name of the village	Distance from new location of Vyahali (in kms.)	Agricultural land obtained	
		in acres	in hectares
1. Asale	2.5	120.00	48.82
2. Bavdhan	4.5	65.36	26.65
3. Kadegaon	1.0	60.31	24.58
4. Songirwadi	10.0	37.35	15.30
5. Pachwad	4.0	24.12	9.81
6. Pasarni	10.0	12.11	4.95
7. Eksar	10.0	4.00	1.61
Total		326.02	131.57

village. The villagers have obtained sufficient settlement area for residential purposes, but the agricultural land given to them is in dispersed patches at different nearby villages with different farms, so most of the villagers have temporarily shifted their residences, near the agricultural farms by constructing small huts on their farms. Such farmers cannot enjoy social life of the village. It is not possible for them to have supervise and to take care of their agricultural crops from Vyahali settlement area. The neighbouring farmers are not giving proper co-operation to the new comers in agricultural operation work. The neighbouring farmers deliberately graze their cattle in the rehabilitated field and by stealing crops. There are two wells within the new settlement one is used to supply drinking water and the other is for irrigation purposes. Roads are not properly constructed within the settlement. No Balutedar (artisan), the Gurav (Temple Pujari), is residing in the village. Land development office of the Wai Panchayat Samittee, has sanctioned Rs. 2000 of loan to each family for constructing residential houses.

At present New Vyahali comes under the Group Gram-Panchayat of Kadegaon Village (Fig. 6.3). Two persons from Vyahali are the members of this group Gram-Panchayat. There is no separate co-operative village society in the village. There is one "Bhajani Mandal" used to sing religious songs at the time of festivals.

A private doctor has been rendering his services of medical help for three hours daily in the village. The villagers are not in favour of the Talathi, who works irregularly and fails to give necessary documents. The villagers wanted to have a separate Gram-Panchayat and a co-operative Society. They are eager to plan a pipe-line water supply scheme for the village. For this purpose they are ready to invest ten per cent of the investment. Nala bunding is essential in the surrounding agricultural area.

There is no S.T. pick-up shed at the stop of Vyahali, on way of Wai-Satara Road.

In the central part of the village, there are Primary School building and the Bhairav temple. Other buildings are constructed newly. There is no Post Office in the village, but letter posted from Bombay takes nearly eight to ten days to reach the village. Villagers are ready to give free land for construction of the Post Office building near the present village Panchayat Office.

The amount contributed in the name of the village deity, is at present with the Kolhapur Charity trust. If this amount is returned to the village, it will be utilised for village development purpose.

There is a crematorium far away to the south of the village, on the right bank of Krishna river. There is no proper road to it. The existing road is full of ditches.

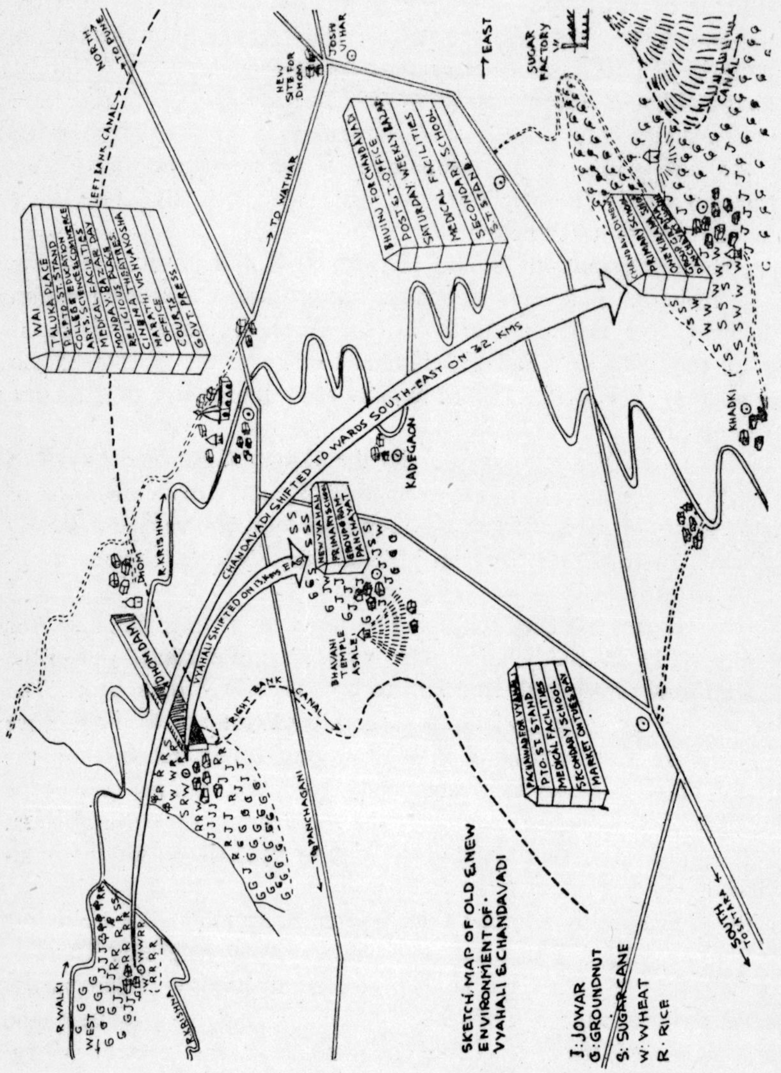

FIG. 6.1 : *Sketch showing Old and New Environments of Vyahali and Chandavadi—their prime crops and their nearby new market centres*

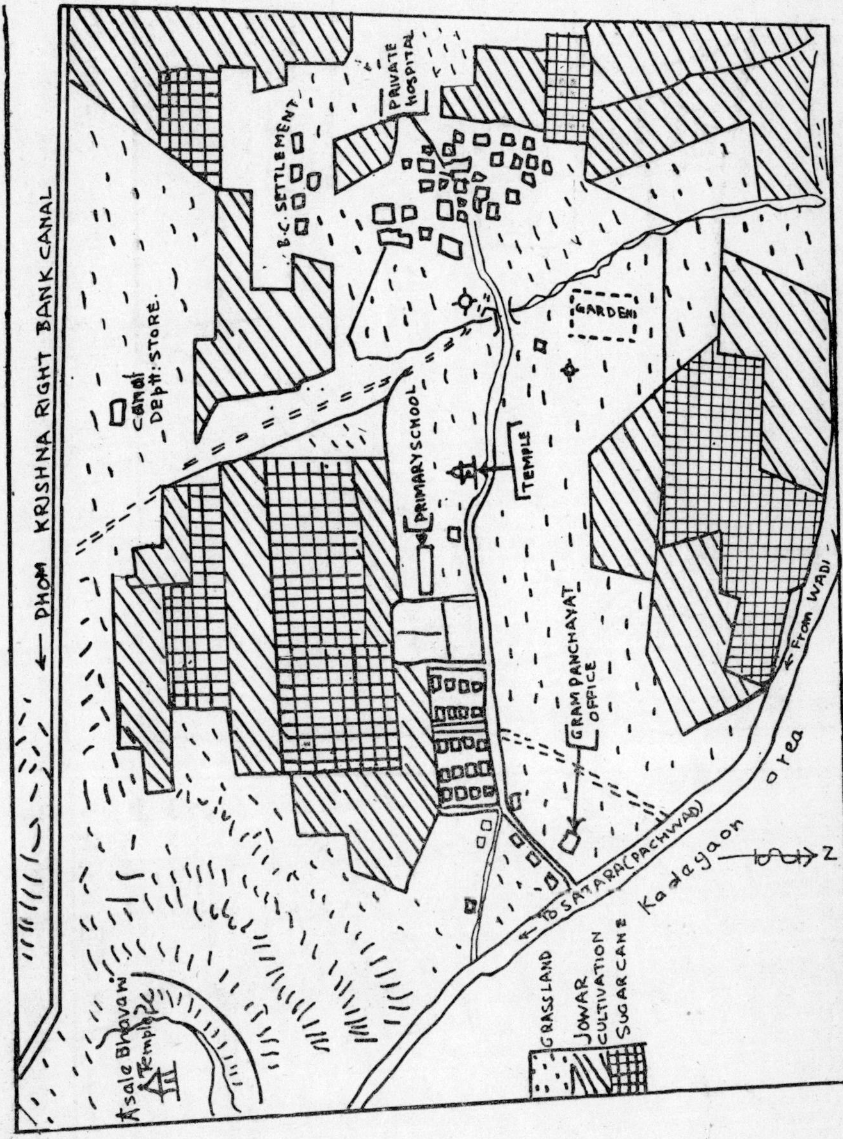

FIG. 6.2 : *Sketch Map of Vyahali*

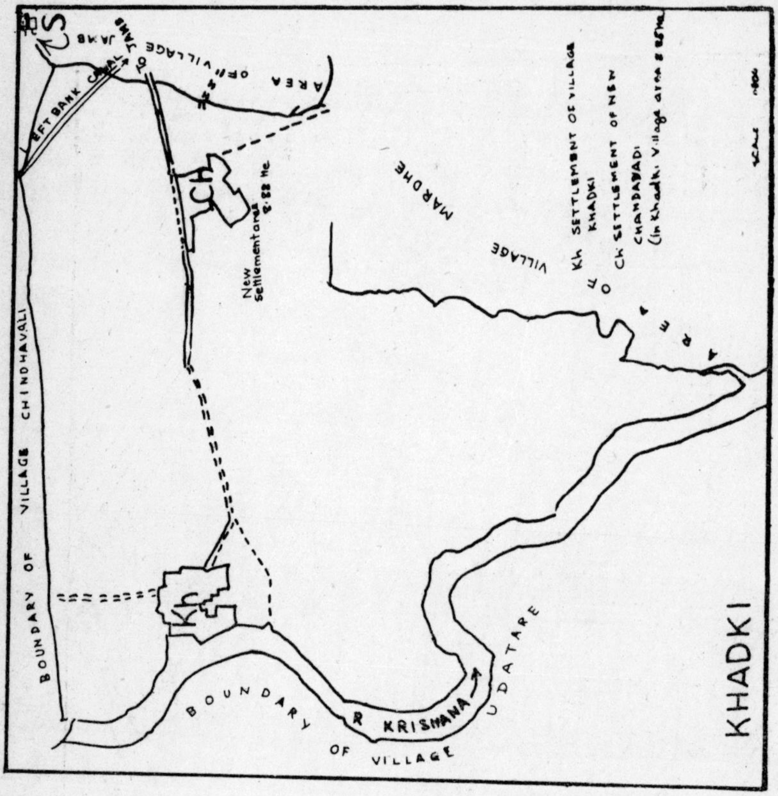

FIG. 6.4 : *New Location of Chandavadi in Khadki Village Area*

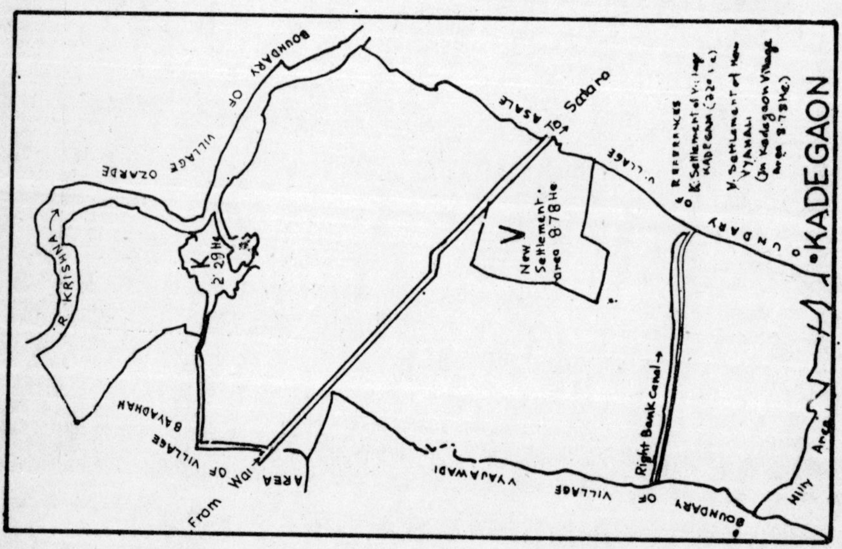

FIG. 6.3 : *New Location of Vyahali in Kadegaon Village Area*

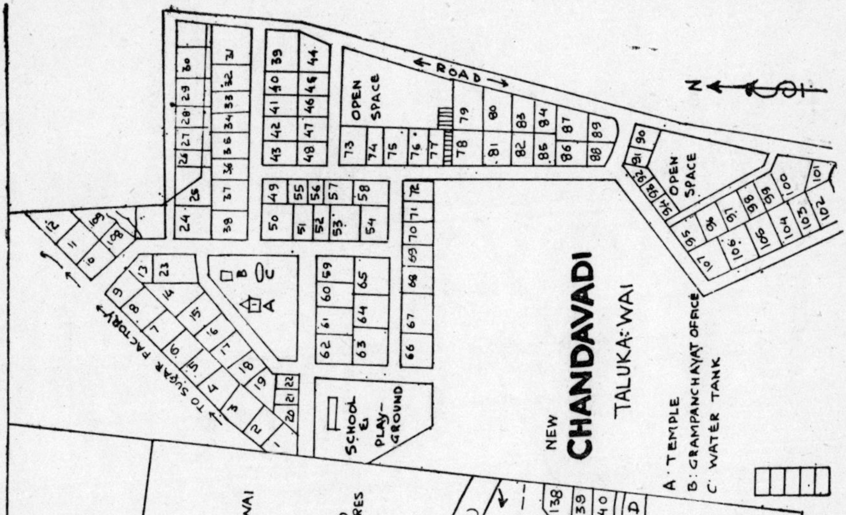

FIG. 6.6 : *Settlement of New Chandavadi (plots)*

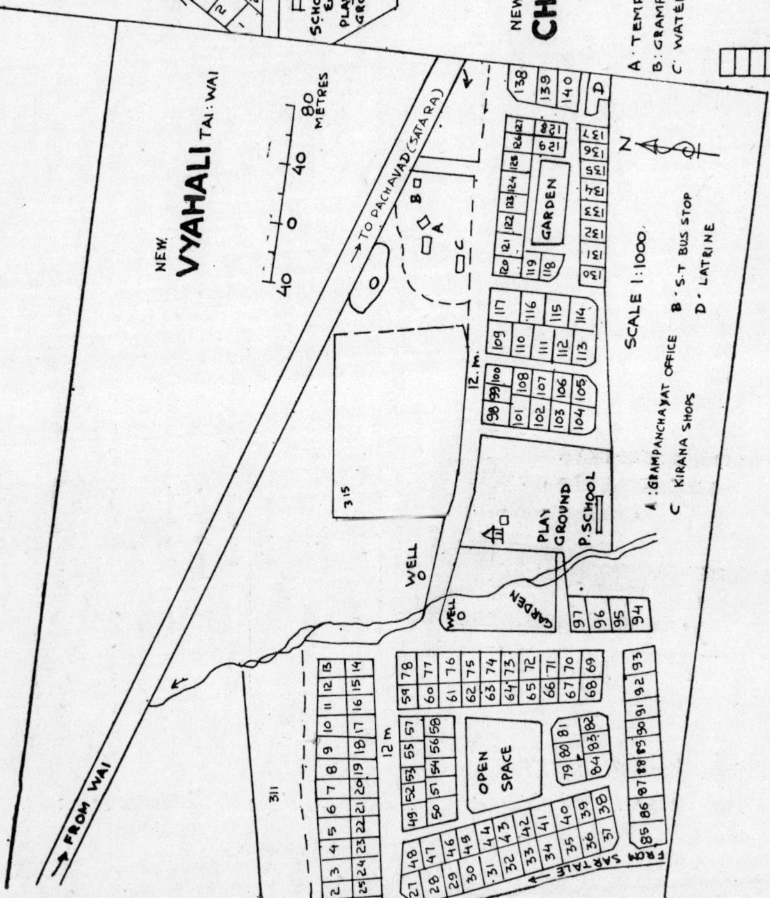

FIG. 6.5 : *Settlement of New Vyahali (plots)*

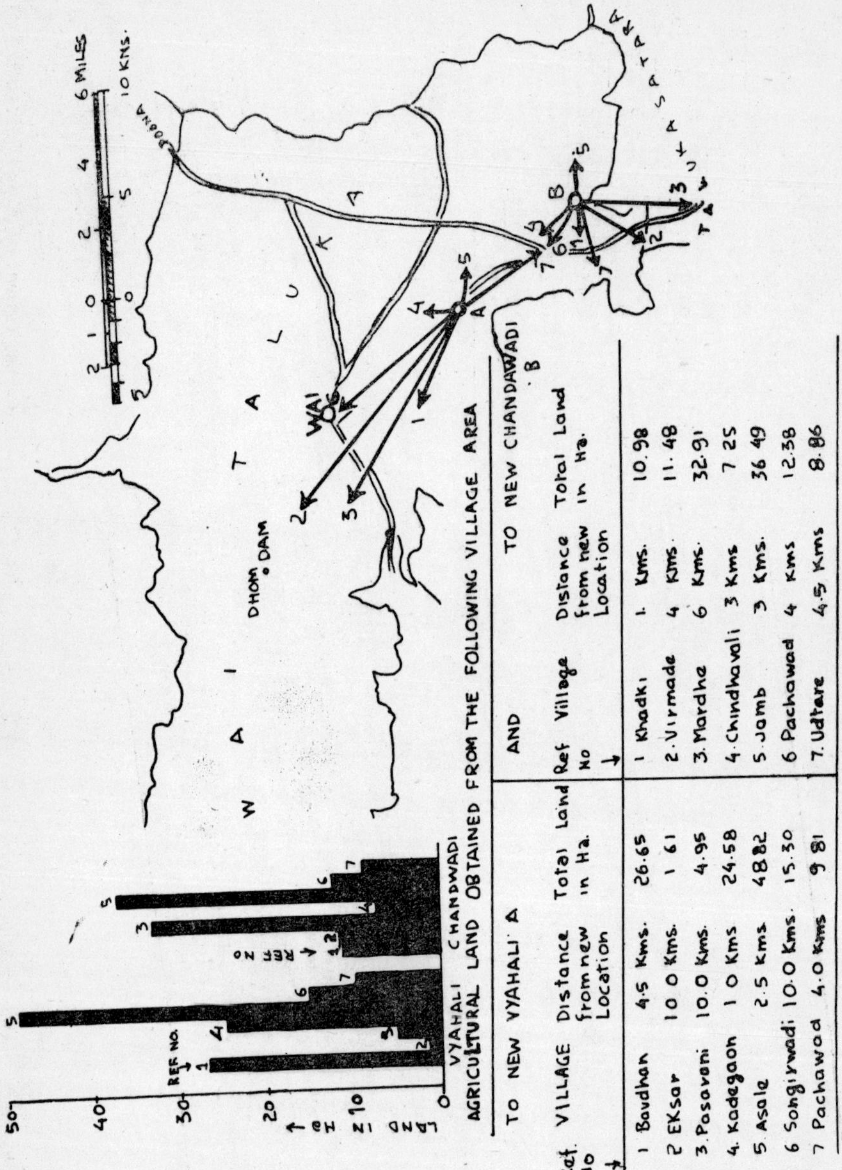

AGRICULTURAL LAND OBTAINED FROM THE FOLLOWING VILLAGE AREA

TO NEW VYAHALI A				AND		TO NEW CHANDAWADI B	
Ref. No	VILLAGE	Distance from new Location	Total Land in Ha.	Ref No	Village	Distance from new Location	Total Land in Ha.
1	Bavdhan	4.5 Kms.	26.65	1	Khadki	1. Kms.	10.98
2	EKsar	10.0 Kms.	1 61	2	Virmade	4 Kms.	11. 48
3	Pasavani	10.0 Kms.	4.95	3	Mardhe	6 Kms.	32.91
4	Kadegaon	1 0 Kms	24.58	4	Chindhavali	3 Kms	7 25
5	Asale	2.5 Kms	48.82	5	Jamb	3 Kms.	36 49
6	Songirwadi	10.0 Kms.	15.30	6	Pachawad	4 Kms	12.38
7	Pachawad	4.0 Kms	9 81	7	Udtare	4.5 Kms	8.86

FIG. 6.7 : *New Vyahali and Chandavadi* :
Agricultural land obtained from nearby villages with distancesi n kms.

The forest department never paid attention to this location for cultivation of trees.

One of the ladies of New Vyahali, said to the investigators, "You are paid for that by the Government to collect our information regarding rehabilitation, but what about us. We do not get anything for providing you necessary informations?" The investigator explained to her as "I come here to study your village. I am not any authority for providing facilities to you. I can help you for writing applications or any other such help which is possible to me".

New Vyahali Through Sample Survey

The studied 20 per cent of families (total 35 sample families) of rehabilitated village Vyahali includes 288 persons (134 Females+154 Males). The average size of the family of population for the universe (14 affected villages) is 6.5 and as per sample it is 8.2 for village Vyahali.

Category "A" up to 1.21 Hectares (3 acres) of Landholders

From the category, 22 households (160 population=82 Males+78 Females) have been studied.

Out of 22, only seven households have obtained 4.56 ha. (12.12 acres) of irrigated land and the remaining households have obtained 9.05 ha. (22.17 acres) of non-irrigated land and the total land obtained is 14.02 ha. (34.29 acres). Per head ratio of land holding comes to 0.85 ha. (8.6 Gunthas).

Table 6.2 shows the villages, at which the above mentioned landholders have obtained agricultural land from the government.

From this category, the per cent of dependents above 16 years of age group is 30 (20% males and 80% females), and women are four times greater than that of men.

TABLE 6.2 : **Name and Distance of Villages from which the Agricultural Land is Obtained to "A" Category of Landholders**

	Village						
	Kadegaon	Asale	Pachwad	Bavdhan	Bavdhan+ Kadegaon+ Asale	Kodegaon+ Asale	Pachwad+ Kadegaon
Distance from New Vyahali in kms.	1.0	2.5	3.0	4.5	4.5+ 1.0+ 0.0	1+ 0	3+1
No. of families	8	9	1	1	1	1	1

Women of Buddhist community work on daily wages as labourers (agricultural).

The percentage of dependents below 16 years of age group is 38 in which 54 per cent are girls and 46 per cent are boys.

Only 31 per cent of families are benefited in rehabilitation. 55 per cent of households have complained about non-co-operation from the village Talathi and Gramsevak.

In the year 1971 and 1972 there were eight families (36% of families) resettled at the new location of Vyahali.

The villagers have obtained their agricultural land during 1971 and 1982. Most of the families (27%) were rehabilitated in 1974, but their agricultural land received the benefit of canal irrigation only from 1976. There was no irrigation facility for five years due to late possession of agricultural land and late availability of canal water.

It is the opinion of most of the small landholders (86% of the total) that they are not properly and sympathetically treated in the process of rehabilitation by the members of the staff of related offices mainly because they are illiterate. They said that they were confused and not guided and even their problems were not considered in earnest by the related offices of rehabilitation and land acquisition.

At the new location only 18 per cent of families have taken interest in tree plantation.

18 per cent of households are disturbed, bifurcated, diversified, their family life due to unstable conditions and due to disputes and quarrels between family members for obtaining the benefit in rehabilitation.

All the studied families (22) have obtained total 16.30 ha. (40.14 acres) of agricultural land. This land comes up to 0.10 ha. (10 Gunthas) per capita. Per family, it comes to 0.73 ha. (1.33 acres). Only 27 per cent of families have obtained residential plots and agricultural land at one location New Vyahali.

Daily per capita availability of milk from dairy farming is 0.05 litres, its utilization per family is 0.4 litre. No single family of this category has started any other occupation from the amount obtained as compensation in rehabilitation. 45 per cent of household received loan from their original village credit co-operative society and nobody has refunded the loan up till now (December 1982).

Out of total families, only 9 per cent of families have established and improved their family life at the new location. 18 per cent of the families are able to maintain their family budgets but others are very poor as they are still fighting against their basic need-food.

It is interesting to see the choice of preferences as given in Table 6.3. The society is represented by Table 6.3.

TABLE 6.3 : **Preferential Vote for the Facilities from the Families of "A" Category**

Preferences for the following facilities	% of number of families agreed to vote for help for the facility
(*i*) Help for construction of a health centre	45%
(*ii*) Help for construction of Temple	31%
(*iii*) Help to the poor sick and disable for medical purposes	22%

Most of all the families (77%), aspired for devotion and worship of "Satayanarayana Puja", and they spend too much over it even by taking loans if they are economically weak. Literacy percentage is 0.6 only.

Fifty-nine per cent of households, have obtained their new agricultural land in time, within two years, but the remaining 41 per cent of families lost five years of waiting time for obtaining possession of agricultural land, from the command area. Only 4.5 per cent of the families have obtained residential plots immediately, but the remaining took two years for obtaining possession of the plots. The 90 per cent of families have not obtained sufficient land as compared to that they had in these original land. They had 7.73 ha. (19.5 acres) infertile land, still unsubmerged and available above the water level of Dhom Lake, whereas they had 4 cows and 11 buffaloes, they have now two cows and 12 buffaloes.

Thirty-two per cent of population is earning population. Out of total population of the studied category of this group, only 13 per cent of population is found employed. Out of these three persons have obtained employment in the Irrigation Department. The remaining ones are agricultural labour workers. Monthly average income of earning population comes per head to Rs. 333.

The studied 22 families were holding total land of 29.13 ha. (72.2 acres) in the original village area, but at present at the new location, they have obtained 14.02 ha. (34.29 acres) of agricultural land, which is one-half of the original land. At the original agricultural area, they were cultivating rice, jowar, and groundnut crops. Now they are in the new area cultivating jowar, wheat, sugarcane and rice. Out of 22, 10 families cultivate sugarcane in their agricultural farms, though they are small landholders.

Out of the total population, 13 per cent are agricultural landholders, 6 per cent are agricultural labourers in which male and female ratio is equal.

One-third of the householders are depending on animal husbandry as a subsidiary income source. They have 20 milk cattle. Milk availability from dairy animals is 0.8 litres to each family.

TABLE 6.4 : **Occupational Structure of the People of "A" Category**

Income source	No. of households depending on it
1. Employments and agriculture	6
2. Labour and Agriculture	8
3. Agriculture and animal husbandry	4
4. Agriculture	2
5. Other occupations	2

TABLE 6.5 : **Yearwise Number of Families Came to New Location from "A" Category**

No. of Households	in %	Year in which they are settled at the new location: Vyahali
1	(12.5)	1971
4	(50.0)	1972
2	(25.0)	1974
1	(12.5)	1977

32 per cent of families have obtained agricultural land connected to the village and one-third of families have obtained good quality of agricultural land according to their expectation. The remaining households have obtained very infertile land which is still (1982) unable to obtain canal irrigation facility. Out of the total households, 36 per cent are B.C. families.

Labour work and services are the main income sources for this category because, they are economically weaker and poor landholders, under dry farming and deficit economic conditions, they have to search out employment or labour work at anywhere (Table 6.4) :

Category "B" 1.21 to 2.02 Hectares (3 to 5 acres) of Landholders

In this category, eight families with total of 93 (51 males+42 females) were studied. Out of the total population the dependents are 31 per cent above 16 years of age group. In it, females are 6 times greater than the male dependents (13% males and 87% females).

Below 16 years the percentage of dependents was 44 per cent. Earning population is 25 per cent of which all are men in majority because women are not allowed to earn except household work. 37.5 per cent families have stated that they received help from the village Talathi and Gramsevak.

12 per cent of households are benefited by rehabilitation.

Persons of this category stand neglected in rehabilitation by the concerned offices.

Only 12.5 per cent of the households have cultivated trees within their new residential area.

Per capita agricultural land comes to 0.13 ha. (13 Gunthas), which is for each family comes to 1.51 ha. (3.30 acres).

Not a single family of this category has obtained agricultural land nearby or close to the residential area.

None of this category has taken any loan from the old village co-operative society.

12 per cent of families are able to maintain within their family budgets. All are traditional minded and give first preference to help for construction of a temple, rather than the construction of a health centre.

Percentage of literacy is 0.04 per cent. Nobody from this category resides within and close to their agricultural farms. 73 per cent of families have to obtain possession of their agricultural land in time. Nearby half of the households have not obtained sufficient agricultural land in the rehabilitated area. Belonging to the eight selected families there is 2.07 ha. (5.5 acres) of grassland area, still available at the old site of the village above the water level of Dhom Dam. Due to long distance, they cannot use it for grazing purposes.

Proportion of dairy animals has been reduced by 15 per cent at the new location.

The studied eight families have obtained agricultural land within different nearby villages as given in Table 6.6.

The above families are holding 0.85 ha. (2.05 acres) irrigated and 11.46 ha. (28.15 acres) unirrigated, total 12.47 ha. (30.35 acres) of agricultural land. Only two sample families, out of eight have obtained fertile land suitable for irrigated crops. Per capita land ratio comes to 0.23 ha. (2.23 Gunthas).

TABLE 6.6 : **Distance and Villages Given their Agricultural Land to the Families of "B" Category**

Village from which the land is obtained	Asale	Kadegaon	Bhuinj	Bavadhan
Distance from New Vyahali in kms.	2.5	1.0	6.0	4.5
No. of families (8)	5	1	1	1

The total population employed in urban area is obtaining per head income of Rs. 506 monthly. All of them are working in the different fields of employments in Bombay.

All the families of this group were holding 14.49 ha. (35 acres) of agricultural land at the old location and they are now at new location, holding 12.47 ha. (30.35 acres) of agricultural land. At the new location in place of rice, jowar, they cultivate sugarcane, rice and groundnuts due to canal irrigation facility. All the families are farmers and nobody of this group works locally as a labourer.

Nobody has obtained agricultural land of their choice. But all of them are taking care of their agricultural land even though it is far away from their settlement location. Due to canal irrigation twenty acres out of 12.47 ha. (30.35 acres) will be irrigated in the near future.

Category "C" 2.02 to 4.04 ha. (5 to 10 acres) of Landholders

Four families have been studied from this category, with total population of 27 (11 females+16 males).

Dependents below 16 years of age group are 33 per cent. The earning population is only 27 per cent of the total.

The families have obtained agricultural land from the nearby villages as given in Table 6.7.

The above four families have obtained 6.87 ha. (17 acres) of total land of which 2.83 ha. (7 acres) are irrigated. Per capita land ratio comes up to 0.25 ha. (25 Gunthas). 8 per cent of population is working in Bombay and from them per head monthly income is calculated to Rs. 450.

These four households, were holding 17.37 ha. (43 acres) of agricultural land at the original village area and now they are holding at the new

TABLE 6.7 : **Agricultural Land Obtained from the Villages with their Distance to the New Village of "C" Category**

Villages	Bavadhan	Pachwad and Asale	Vyahali (Kadegaon)	Udatare
Distance from New Vyahali (in kms.)	4.5	3.0 & 2.5	1	10
No. of families obtained the land (4)	1	1	1	1

TABLE 6.8 : **Occupational Structure of the Landholders of "C" Category**

Occupations	Agriculture	Agriculture and services of employed persons	Agriculture and animal husbandry
Number of families engaged in it (4)	1	1	2

village area only 6.87 ha. (17 acres) of agricultural land which is 2.5 times less than that of original.

At the old location, they were obtaining production of rice, jowar, and ragi agricultural crops. Now at the new village area, they started to cultivate sugarcane and turmeric new crops in their farms. Nobody of this category, is found engaged in labour work at the other's farm.

50 per cent of families have dairy farming as a subsidiary occupation.

Only 25 per cent of households (one family), have obtained agricultural land as per their choice.

Year	Families resettled in the New Vyahali (No. of families)
1972	3
1974	1

Nobody of this group is found taking interest in cultivation of different trees within the new area.

Agriculture land comes per capita 0.25 ha. (25 Gunthas) and for per family 1.71 ha. (4.10 acres).

The affected and unemployed persons of this group have never tried to take benefit of getting facilities of jobs in irrigation department. Nobody is under the pressure of loan taken from the old village society. 25 per cent families are able to maintain family budgets. The remaining 75 per cent of families are unable to maintain this budget. Nobody is settled within his agricultural farm. 75 per cent of families have acquired their agricultural land very easily.

Category "D" Landholders of 4.04 to 6.06 ha. (10 to 15 acres) of Landholdings

Only one family has been studied under this group. (5 males+3 females.

Total 8 persons). Dependents above 16 years of age group are 25 per cent and below 16 years of age group are 4. The earning persons are 2.

To this family, the agricultural land obtained to the very close of the village settlement area, from which one acre is cultivated under irrigated crops, and remaining 5.65 ha. (14 acres) are under non-irrigated crops. Agriculture is the only source of income and nobody is employed elsewhere in urban centres. In original village jowar, rice, cereals and small quantity of sugarcane were the main crops, where the family was holding 7.27 ha. (18 acres) of agricultural land. Now the same family cultivated the same crops including turmeric and vegetables at the new obtained agricultural farms.

Livestock activity with agriculture is an important supporting activity of the family. From dairy animals, milk production ratio comes to 1 litre daily. In 1972, the family, came to settle at New Vyahali.

Per capita agricultural land comes to 0.75 ha. (1.35 acres) which is highest in the village. This rich family is found against investments on showy functions and on traditional and unprofitable functions. All the households of rich groups have obtained agricultural and residential lands in time.

The strength of dairy animals has been reduced at the new location due to unavailability of grassland area at the new village.

Village Vyahali : Study of all the Selected Families

35 sample families (32% of families of the village) out of 108 total families have been studied from the village. Within the above families, there are 25 per cent of dependents (17 per cent males+83 per cent females) above the age group of 16 years. For the village, the dependents of female are six times more than the male dependents. The dependents below the 16 years of age group are 50 per cent to the total population (55% males+45% females) (Fig. 6.8).

In Vyahali, total earning population is only 25 per cent in it, women are 15 times smaller than that of men.

The selected 35 families have obtained 39.50 ha. (97.24 acres) of agricultural land of which 9.25 ha. (22.32 acres) are under irrigated and 30.25 ha. (74.32 acres) are under non-irrigated crops at present. Per capita land comes to 0.14 ha. (14 Gunthas) and 0.88 ha. (2.8 acres) for each family (Fig. 6.6).

At present not a single lady member of the village is found as an employee working in public or private services.

The employed persons working in different services at urban centres are obtaining monthly per capita income of Rs. 515.

At the original village area, selected 35 families were holding total 74.99 ha. (168.2 acres) of agricultural land and from it they were cultivating

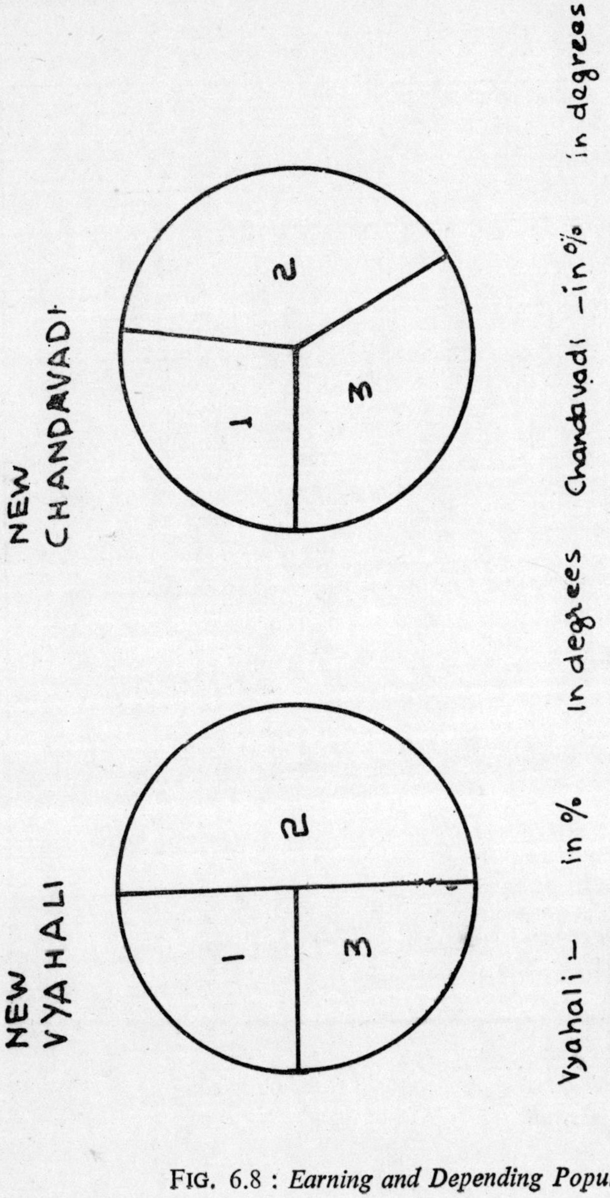

FIG. 6.8 : *Earning and Depending Population*

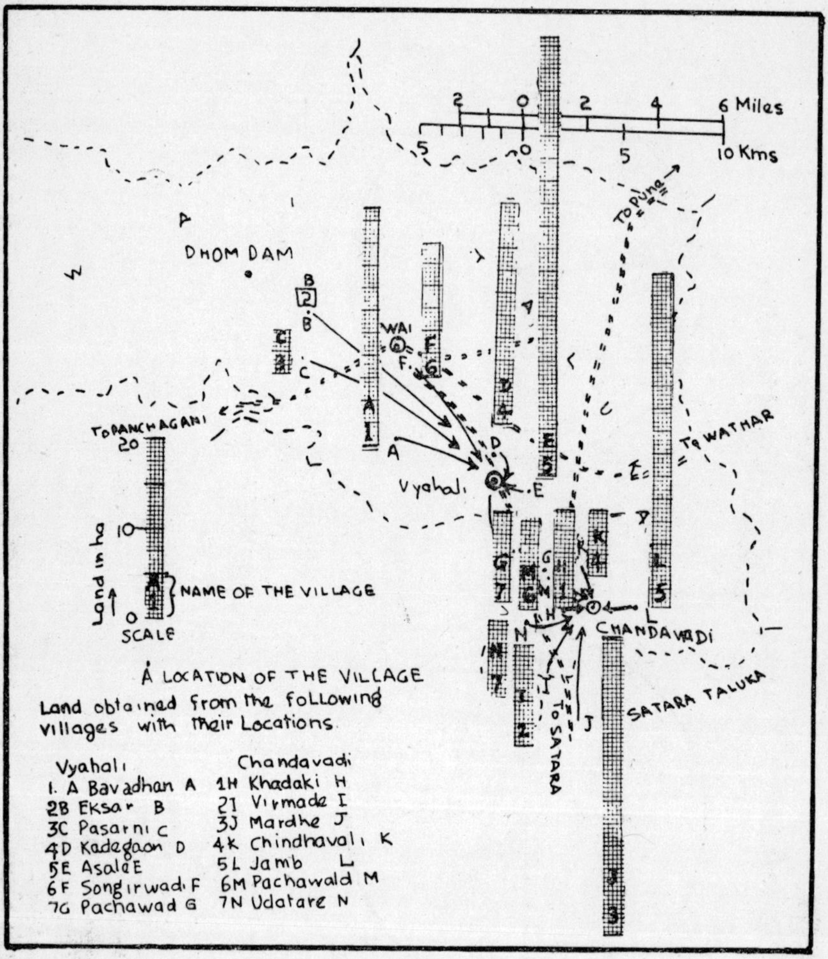

FIG. 6.9 : *Bar-Graph showing Agricultural Land Obtained to New Vyahali and Chandavadi*

jowar, bajra, wheat, groundnut and rice crops. Now the same number of households have obtained only 39.50 ha. (97.24 acres) of agricultural land which is half of original agricultural land.

In the new village, 17 per cent are the cultivators and 4.3 per cent are agricultural labourers working on daily wages.

The strength of male and female labourers is equal. This is only because on an average, most of the families are poor and shifting have made them to work as "labourers", since no alternative job is available to them.

45 per cent of the families are supported by dairy farming as a subsidiary occupation, but no scientific dairy farming is developed in village. The animals are grazed along the mountain slopes to the backside of the new village, on rainfed grasslands. With the population, per capita milk consumption is only 0.01 litre.

Nearly 14 per cent of families of the village are holding agricultural land closed to the new village. 77 per cent of families of the village have not obtained suitable agricultural land and they are not satisfactory with their rehabilitation position. Daily milk supply from dairy animals, comes up to 23 litres for 35 sample families.

It is also experience of 77 per cent of families of New Vyahali that with them the agricultural neighbourers from other nearby villages are not giving co-operation in the agricultural implementation work, three farmers have explained that protection of fully grown crop from the thieves became very difficult and impossible due to more distance between the settlement and agricultural land obtained in rehabilitation (Fig. 6.7).

For the present (1982-83), 9.20 ha. (22.32 acres) of land is benefited by canal irrigation. Now within two to three years, after construction of channel routes up to the agricultural farms, more land will come under canal irrigation.

Only 1.3 per cent of persons have taken benefit of employment as dam affected persons in irrigation and rehabilitation departments. The villagers have started to adjust within the new rehabilitated environment.

They remembered the days of past life of the old village. They were happy and natural vegetation of the valley was an important source of income to them. Jambhul, phanas, mango, karvand, teak, babul were the important trees of the area.

Out of the studied sample families of the village, 60 per cent of families are against village Talathi and Gramsevak due to improper guidance.

The economy of 68 per cent of households suffered loss in rehabilitation. They do not like the new location of the village. They were compelled to shift to the new location. They were helpless.

They said that the amount of Rs. 5000 to each family, sanctioned by the government for construction purposes was not sufficient.

TABLE 6.9 : **Per cent of Families Coming to New Vyahali and Benefited by Canal Irrigation**

Per cent of families (in %)	Number of studied families	Year in which canal irrigation facility becomes available to them	Total number of years invested for waiting of canal irrigation facility—unproductive period of irrigated crops from 1974
5	2	1979	5
5	2	1978	4
40	14	1977	3
2.5	1	1982	8
20	6	1976	2
25	9	1980	6
2.5	1	1981	7
100	35	—	—

TABLE 6.10 : **Yearwise Number of Families Settled at New Vyahali**

Year	Number of selected studied families settled at New Vyahali (35)
1971	9
1972	22
1973	—
1974	3
1977	1

TABLE 6.11 : **Period for Obtaining Possession of Agricultural Land and Facilities of Canal Irrigation for New Vyahali**

Number of families (A)	Waiting period in years for obtaining possession of agricultural land and facility of canal irrigation (B)
2	3
4	4
12	5
4	6
4	8
6	9
1	0
1	7
1	10

From Tables 6.9 and 6.10 above maximum number of families have lost 3 years of unproductive helpless period at the new settled plots by constructing shelter and working as labourer at the nearby villages.

After settling at the new location, there are maximum number of families who have lost their 5 years of period as non-productive and as waiting period of possession of agricultural land and canal irrigation facility (Table 6.11).

We can say to this period of five years as an "exile period" to the dispersed affected persons by Dhom Dam. In this period, there was no stability, and no source of income facility and no peace to their mind.

Only 14 per cent of households, have taken interest of cultivating trees within their plots, and 17 per cent of families obtained land closed to the new settlement location and 5 per cent of families have improved their economy.

Up to 2.02 ha. (5 acres) of landholders, from the category A and B are found, that they have concentrated for keeping dairy animals as a subsidiary support to their income, but this occupation is practised by old traditional methods.

57 per cent of small landholders up to 1.21 ha. (3 acres) of landholding have disturbed their family life by rehabilitation.

31 per cent of families have still accepted of taking dowery in the marriages of their sons as a "gift".

31 per cent of affected households, from rich group of landholdings were supported by economic help from their nearest relatives even in fixing the marriages, the relatives have maintained good relation with these affected households during the unsheltered position. The remaining poor and medium dam affected landholder (69 %) are observed very serious due to unstable life about their new environment, and because their friends and nearest relatives have stopped their co-operative approach with them. They become now as "unsheltered".

17 per cent of families are able to maintain their family budgets, and the remaining are depressed under poverty.

The villagers are giving more importance to "Satyanarayan Puja" rather than the annual fair "Jatra".

The strength of educated persons is 3.9 per cent (81% are S.S.C., 9% are H.S.C. and 10% are graduates). No Post-Graduate person of any faculty is observed in the village.

Out of 35 sample families, maximum number of families are under "A" category, having below 1.21 ha. (3 acres) of landholdings. They are still the owner of 7.88 ha. (19.20 acres) of available infertile land at the old location in the upper part of the valley, above the water level of Dhom Dam.

The dairy animals are reduced by 18 per cent in the new location to that of dairy animal from original location.

In June 1982, at New Vyahali, Shri Pol explained that he has obtained 2.13 ha. (5 acres, 11 Gunthas) of agricultural land from Bhuinj village, (under No. 25/44 of village abstract of Bhuinj). But the original landholder is not ready to leave his possession of the land though the land has been acquired by the Collector of Satara. Shri Pol of Dom village have lost along with his brother 24.27 ha. (60 acres) of agricultural land under Dhom Project.

It is observed that no houseship and emotional entanglement have been formed within the new locations from the nearby villages of Vyajwadi and Asale. (For the available facilities to the old and new locations of village Vyahali please see Table 7.7 in Chapter 7).

7

New Chandavadi

New Settlement

THE village is newly (1975) settled at a distance of 20 kms. south-east from Wai, near Bhuinj Sugar Factory (The Satara Sahakari Sakhar Karkhana Ltd. Bhuinj) within the Khadki village (Tal. Wai) area at Jamb Mal. (Fig. 6.1 and 6.4). The distance from original to new location is 32 kms. Bhuinj and Pachawad are the two market centres located nearly 5 to 7 kms. away from the new location.

There are sufficient medical, health facilities, stationery and cotton cloth shops, and secondary and higher secondary school facilities in the market centres. Tuesday is the weekly market day for Pachawad. Pachawad is important market centre in the area and at the time of annual fair, "annual cattle bazar" is an attraction to the villagers of the region. At Pachawad and Bhuinj there are no cinema theatres and other recreation facilities.

At new Chandavadi government have resettled 96 families including eight B.C. families. Old Chandavadi and old half part of village Velang shifted at Khadki village area, formed the settlement of New Chandavadi. (Fig. 6.1 and 6.4). Khadki is a gram-panchayat centre as well as post office centre to New Chandavadi. At New Chandavadi there is a primary school with 136 students, but for Secondary and Higher Secondary Education, the students have to go to Bhuinj or Pachawad. For college education Wai is the only nearest centre to them.

The total area occupied by the village settlement is 8.88 ha. (22 acres) (under Khadki village of command area in Wai Taluka), the total agricultural area is 101.17 ha. (250 acres). Nearly all the land comes under canal irrigation. The agricultural land is obtained by new Chandavadi from the nearby villages as given in Table 7.1 (Fig. 6.7 and 6.9).

TABLE 7.1 : **Villages with their Distances from which Agricultural Land Obtained to New Chandavadi (Fig. 6.7)**

Name of the nearby village	Distance from new location in kms.	From Wai	Agricultural land obtained	
			in acres	in hectares
Khadaki (Chandavadi)	1	18	27.07	10.98
Viramade	4	20	28.17	11.48
Mardhe	6	22	81.14	32.91
Chindahavali	3	12	17.38	7.25
Jamb	3	19	70.10	36.49
Pachawad	4	10	30.26	12.38
Udatare	4.5	13	21.38	8.86
Total			277.30	

TABLE 7.2 : **The Strength of the Primary School of New Chandavadi (1982-83)**

Class	Boys	Girls	Total	B.C. Boys	B.C. Girls
I	7	5	12	2	1
II	7	5	12	3	—
III	12	15	27	1	1
IV	10	12	22	1	3
V	17	11	28	1	1
VI	17	5	22	2	—
VII	10	3	13	1	—
Total	80	56	136	11	6

The total population of the new village is 500 (260 males+240 Females).

There are four teachers in the primary school (3 males+1 female).

Settlement of Chandavadi is divided in two parts. The northern settlement is covered by the affected persons of old Chandavadi and the southern part is occupied by the affected village Velang. (Fig. 6.6) (Village Velang is the biggest affected village by Dhom Project and partly resettled within the New Chandavadi).

The new settlement, is connected with Wai, (Taluka place), by S.T. Bus route, on this route are located Bhuinj village and "Satara Sahakari Sakhar Karkhana".

Within the residential area connected cattle sheds have been constructed by many of the villagers. Most of the houses are roofed with tin sheets.

The temple of the village is very simple, constructed by tin sheets as a temporary hut, and all the idols of gods are placed in it. Three cement

Water Tanks have been constructed but due to incomplete work of pipe line connections, they are not in use for the present (Feb. 1982). Supply of drinking water is provided by rented water tankers and its annual charges are collected by the villagers. Canal is passing from the eastern part of the settlement. Canal water (Left bank Krishna Dhom Canal) is used for the purposes of bathing and washing their clothes. The village is connected with Bhuinj Sugar Factory (Satara Sahakari Sakhar Karkhana Ltd., Kisan Veer Nagar) by very rough road, which is not motorable in rainy season.

The track, used for transporting sugarcane, connected the village with sugar factory, becomes very rough. Within the settlement there are rough roads and within the area of residential plots, vad, pimpal, mango trees are cultivated by many farmers.

It is to be noticed that there are 60 per cent of families from which women are also helping to men in the work of agricultural farms.

Difficulties in Leaving the Old Location

The people were not informed about the details of evacuation. Many thought that their village would never be submerged. It is habitual to them that government gives many plans but never implements quickly. So they do not take any announcement of leaving their village seriously. In the month of July 1976, due to a heavy rainfall, the water level of Dhom Dam started increasing and encroaching over the settlement area of Chandavadi and no sufficient time is given even to take and remove and transport their residential house materials towards the new location. No information was given to them about "to shift" immediately. Due to rainy climate, people suffered from cold. They observed before their eyes, due to high speed of increasing water level of the dam, their houses, temples, big trees, were submerged by the rising water. This was the last time for them to see with their eyes the picture of their original site, encroached and covered by the dam water, surrounded the tombs of their temples, roofs of their houses. They lost their beautiful original environment forever. They said goodbye to their original environment. This calamity of increasing dam water level, was observed by the villagers, by ascending up to the backward mountain spur. The roaring water waves of the dam captured the agony of the village. The big statues and carved sculpture of big temples were lost under the water.

The village area was once covered with dense forest, and with rice and jowar cultivation. Now natural beauty was lost under the lake. No relic remains. The villagers are the only animate relic remnant of the old village. The old village is recorded in the past history of the nature.

It was raining heavily and in this situation they shifted and came to the new location. This shifting was like a dream to the children. For the youngsters and elder persons, it was unbearable. To the old persons, who spent their major life in the old environment, became frustrated, and uneasy due to their displacement. There was mental pressure as they were unstable. It was a great challenge to them for utilizing their remaining period of life for building up the new environment. To wait for and carry the drinking water provided by water tankers is an important work in their daily life.

People of Chandavadi spent 72 hours in rain by the time they came to the new location, where there was no shelter for them. Many of them were sick under cough, cold and fever diseases. At the beginning, 'rehabilitation department' have been constructed temporary shade in which only 10 to 12 families were sheltered. The remaining families were living under the roof of sky and wet rocky ground below. Some of them have constructed temporary shelters known locally as "khop". The cows and buffaloes, sheep, started crying under the unroofed new environment. Some of them have sent their cattle to their relatives at nearby villages. At the old location Shri. D. Pol maintained a garden including 60 mango and 4 Phanas trees. This beautiful garden he lost under the project. Shifting affected his frustrated mind so severely that he died due to attack of high blood pressure. A family originally holding 2.02 ha. (5 acres) of agricultural land, is now still landless in the new location and members of the family are working as labourers in the agricultural fields in nearby villages on daily wages.

The old village was a beautiful pleasant village. Summer was very pleasant season of this area. Now, the new village seems to them unpleasant at all the seasons.

A family, of village Boriv is now settled at new Chandavadi alone. The remaining all landholders of Boriv, are shifted at Rahimatpur. Due to longer distance from Chandavadi to Rahimatpur (Tal. Koregaon), the family could not take benefit in social work and could not keep contact with the relatives specially at the time of festivals and marriages and other social functions.

The family has to develop new contacts with their neighbours. Water for drinking purposes is (1982-83) distributed with the help of tankers. Many of the family members are working as an agricultural labour work in the nearby villages on daily wages. It becomes difficult for them and also it is not possible to kill the waiting period for arrival of water tanker, which comes in the afternoon time and that too in limited quantity.

The land given to the affected landholders is infertile and unlevelled. Still no water channels have been constructed properly within their agricultural farms.

Shri G. Pol, has still (1982 Sept.) not obtained agricultural land at

Chandavadi area. He has taken the benefit of getting employment in irrigation department being a dam affected person.

Smt. Parvati Shinde, was holding 2.42 ha. (6 acres) of agricultural land at old location. Now she has obtained only 0.45 ha. (1.5 acres) of infertile land. The land is not levelled and for herself, it becomes difficult to cultivate irrigated crops in it, though water is available to the land. To improve the land, majority of landholders should have to make improvement of agricultural land (i.e. levelling, bunding the farms).

Smt. Parvatibai, possessed her land property after fighting against the obstructions raised by the original landholder, and for this work, she spent, 3 years of period, for which she lost her too much energy, time and money.

Due to poor economic conditions they are helpless to maintain their farms. To take care of such agricultural farms, which are far away from their residential village, is also another problem. To keep watchful eye on cultivated crops, becomes very difficult for them.

It is the demand of many landholders to provide work for them by generating agro-based occupations. Fertilizers, seeds and other equipments should be provided to them by co-operative small units. The government should also provide medical facilities, lavatory and sanitary blocks to their settlements.

In rehabilitation, a landholder has obtained agricultural land from the villages Mardhe and Pachawad, but no possession of the land from Pachawad village is acquired by the landholder due to raised objections and obstacles from original landholder.

Shri. G. Shinde came at the new location in 1978, but up to (Dec. 1982), he could not obtain possession of his agricultural land and he is waiting still for the orders of "Rehabilitation Department".

Due to drinking of wine one affected person has lost all the amount obtained in compensation by rehabilitation. Now his son is trying to rebuilt his family economy by doing agricultural labour work.

Shri Kakade, (from affected village Velang) originally was holding 4 cows and 6 buffaloes, but now due to unavailability of grazing land, at new Chandavadi, he sold his cattle at the market centre Pachawad.

While construction work of the dam was under progress Chandavadi villagers were at the original location. After completion of the project work, in June 1976, is the first rainy season and the first time that water stored in the lake started to increase its level.

At the end of June, the level of the water reaches upto the settlement area of the village where the encroachment of water could not give sufficient time to the villagers to shift elsewhere at protected area, with all of their house materials. It was raining, and the people started to shift with the help of trucks provided by irrigation department and came into the new location at Khadki Tal. Wai at Jamb Mal. It was the most painful unprotected period of their life. Many villagers were captured by coldness and fever diseases. The velocity of the wind was very high. They could not protect their cattles. They are like the dumb driven creatures.

One of the old persons said that, "shifting from old to new location" feels him as if leaving the shelter of love of his father and mother. The relationship with the nature is converted in the new environment, which brings about as if overall changes in the life of a bride, when she enters in the groom's home. She accepts new way of life and accepts the life of new landscape.

The land obtained from Khadki area, is very infertile, and unlevelled. Big landholders have obtained fertile agricultural land. They are happy in the new settled area.

Some of the big landholders of new Chandavadi made complaint about the sugar factory as they have cultivated sugarcane in their agricultural land, but no preference is given for their sugarcane to make sugar in the factory. The Satara Sahakari Sakhar Karkhana (Bhuinj) is giving first preference to those who are the share holders of the factory. In the affected area, the majority of villagers were not cultivating sugarcane. They are not share-holders of the factory. Due to this reason, though they have cultivated sugarcane, they lost the claim, and obtained last chance at a very late period, resulted less yield of sugar from their sugarcane.

The sugar factory site is very near to New Chandavadi. (Having 4 to 5 kms. distance).

In the year 1982-83 nearly 150 households (families) with their bullock carts, (one bullock cart of each family), from Bhir district of Maharashtra State, arrived and settled outside the sugar factory area. They were helpful for transportation of sugarcane from agricultural fields to the factory in cheap rates. No local source of manpower is observed engaged in this transport work of sugarcane by bullock-carts.

A religious minded person, who denied to give information regarding his family, and given non co-operation with us. The person was depressed by the loss of land under Dhom Project.

Now the settlement of Chandavadi is coming up, slowly and many of villagers have construction work of house. Within the New Chandavadi village area, in the exile period of 2 to 3 years which is non-productive period, nobody from society of benefited area have supported and given any type of help to them.

At the original village area, some landholders were supported by subsidiary source of income from mango, jambhul and chinch (Termarindus Indica), Phanas trees. Now they have to wait and spent 15 to 20 years of period to obtain income from economically important trees, which they have cultivated at the new area.

Not a single photograph was available with the villagers about their original environment. Food for all family members, was a problem before them, because 75 per cent of the landholders were small landholders, coming under poor economic position. Due to poverty it is not possible to enjoy by investing amount for taking photographs.

The pleasant environment and climate of the original location in the upper Krishna Valley of Wai Taluka was very close to the plateau of Panchagani hill station. Now they have come in such area where there are no perennial wells, but barren lands.

The New Chandavadi : A Sample Survey

Out of 92 landholders of new location there are 38 families (41%) of households) studied from the village of new Chandavadi, in which two families are included from B.C. communities. (One from Buddhist and one from Mang community). Within 38 families, there are 252 persons (120 females + 132 males). Average size of the family of population for universe (for 14 fully affected villages) is 6.5 and for Chandavadi also it is the same.

Category "A" up to 1.21 ha. (3 acres) of Landholders

The total number of sample families selected for study are 23, includes 126 persons (59 females+67 males). Out of that, a family was originally holding 0.85 ha. (2.5 acres) now remained landless and waiting for agricultural land from the government. Due to unavailability of agricultural land, there are 2.6 per cent dam affected landless households are on the waiting list of the government.

The remaining 22 families have obtained agricultural land from the government, in villages as shown below :

Name of Village	Khadki	Mardhe	Mardhe & Pachawad	Virmade & Jamb	Khadki & Virmade	Khadki & Jamb	Khadki & Mardhe	Jamb & Mardhe	Khadki & Jamb & Udature
Number of families obtained agril. land	8	2	2	1	1	1	4	2	1

Affected families of B.C. community have still not obtained possession of agricultural land from Pachawad and Mardhe area, because they are obstructed by the original owners, though the land is acquired by the government.

The dependents above 16 years of age group are 25 per cent (20% males+80% females) and the dependents below 16 years of age group are 40 per cent (52% boys+48% girls) to the total population.

Earning population of this category is 35 per cent with the total population in which men are five times greater than the women. (males 83% and females 17%). All the families of this group are holding total 19.65 ha. (48.22 acres) of agricultural land in which 6.21 ha. (15.15 acres) are under irrigated crops and 13.40 ha. (33.7 acres) are under non-irrigated crops. Per capita land ratio comes to 0.16 ha. (16.1 Gunthas). Not a single woman is employed elsewhere.

Earning population employed at Bombay and Indore Cotton Mills, as workers, are 12.5 per cent. There is a dam affected primary teacher, working in the school of his own village New Chandavadi. Average per capita income of the employed population is Rs. 536 monthly.

Total Number of families	Holding land at old village	Holding land at rehabilitated new village
23	120.16 acres (48.71 ha.)	48.22 acres (19.61 ha.)

The 23 households were holding at old location 48.71 ha. (120.16 acres) of agricultural land in which they were cultivating rice, jowar, sugar-cane groundnut, ragi, vari, nachani and cereals. Now they have obtained only 18.32 ha. (45.14 acres) of agricultural land which is thrice less than that of the original. Due to this it is clear that the affected villagers have obtained at new area insufficient land. Nearly 75 per cent of the land obtained in rehabilitation is infertile and barren.

In this category, the cultivators are only 18 per cent (in which 81% are men and 19% are women) to the total population.

The labourers are 50 per cent (in which 63% are men and 37% are women) to the total population.

Preferential occupational structure of the studied families is given below:—

Occupation	Number of families
1. Agriculture and Labour work	7
2. Agriculture and services (employed elsewhere)	7
3. Services and agriculture	4
4. Agriculture	2
5. Services	1
6. Labour work	1
7. Services, Labour work and Agriculture	1
Total	23

Compared to the milk production from dairy animals of this group, per person daily milk consumption is only 0.2 litre, which is one litre to each family. On an average, 23 per cent families have obtained agricultural land closed to the village settlement. 90 per cent of families have obtained good type of agricultural land. Out of total 18.32 ha. (45.14 acres) of obtained land 15.09 ha. (37.14 acres) are supported by canal irrigation facility. Affected persons are very easily mixed with their new environment and are called by the villagers of command area as "unsheltered guest families", locally known in Marathi as "Upara".

The villagers of command area are not ready to help to the displaced persons and to accept the view that they are benefited by Dhom Project which is constructed on the lands of displaced families.

66 per cent of affected families have explained about non-co-operation by the landholders from command area. 30 per cent of families of this category, have explained about non-co-operation of help from the village Talathi and Gramsevak about agricultural and rehabilitation work. The new location is actually disliked to 91 per cent of the villagers but due to no other source for them, they prepared themselves compulsorily adjust with the new environment. 74 per cent of the households feels rehabilitation non-benefited to them.

Year	Per cent of number Studied families resettled at the new location
1975	10
1976	69
1977	13
1978	4
1979	4

48 per cent of households have taken interest in cultivation of trees in their village plots. 8 per cent of families have been disturbed by their family members who have mis-used the amount obtained in rehabilitation to them. Per capita agricultural land ratio comes to 14 Gunthas (0.14 ha.) and to each family it comes up to 0.76 ha. (1.36 acres). The size of the average residential plot distributed to each family is 0.38 ha. (3.8 Gunthas) (per head only 0.07 ha. (0.7 Gunthas).

Only 0.03 per cent of persons of this group have obtained employments being dam affected persons in irrigation and rehabilitation department. The Wada structure of original family life is disturbed in the new settlement. Only 4 per cent of people are benefited to settle closed to each other at the nearby plots. All the households (35%) were taken loans from the original village society and all of them have returned the amount of loans and now they are free for thinking to form a new co-operative village society at the new location.

Due to traditional impact, 30 per cent of families are very backward and are ready to accept dowery system, which never feels to them as unhuman.

69 per cent of families are poor and are found difficult to adjust their family budgets and accounts.

44 per cent of households have not obtained the possession of agricultural and 4 per cent of households, the possession of residential plots in time.

74 per cent of households have not obtained sufficient agricultural land at new location. Only 9 per cent of families are found properly rehabilitated.

In catchment area at old location, above the water level of the dam, 9.10 ha. (22.22 acres) of unirrigated land is available from the studied families of this category, which is used for jowar cultivation.

The dairy animals have been reduced by 48 per cent for the present to that of original village, due to unavailability of grazing land in the new location.

Majority of the landholders have given first preference for construction of health centre and temple, rather than helping to the physically handicapped persons of their village.

The ratio of unemployment is 0.06 per cent (0.02 female and 0.04 male) to the total population. The educated persons are only 11 per cent.

Category "B" Landholder of 1.21 to 2.02 ha. (3 to 5 acres) of Landholdings

Sample families with 82 population (39 males+43 females) are studied from this category. The dependents, above 16 years of age group are 39 per cent (50% boys and 50% girls).

The earning population is 32 per cent (19% females and 81% males).

For the 10 households, the agricultural land obtained in rehabilitation is 16.12 ha. (39.36 acres) in which 9.19 ha. (22.31 acres) are used for the present under-irrigated and 6.93 ha. (17.5 acres) are used for non-irrigated crops. Per capita agricultural land comes to 0.19 ha. (19 Gunthas) and to each family it is 1.51 ha. (3.36 acres).

Village from which the agricultural land obtained	Pachawad	Pachawad & Khadki	Khadki	Khadki & Jamb	Khadki & Mardhe	Jamb & Mardhe	Jamb & Mardhe & Udatare
Number of families studied (10)	1	1	2	1	2	2	1

The employed population in different services, (working at Bhuinj, Wai and Bombay) is 9 per cent and by it the monthly per capita income is Rs. 250.

From the original agricultural land, they were obtaining rice, cereals, turmeric, wheat, jowar, groundnut and very small quantity of sugarcane and vegetables. For the present, in the new agricultural area, the same cropping pattern (excluding turmeric) is developed including 'ginger' a new crop to them.

In the old village area, they were holding 79.03 acres and now they have obtained only 16.12 ha. (39.36 acres) of agricultural land, which is nearly half to that of old land. The strength of the cultivator is 12.2 per cent. The labourers are 7.3 per cent (labourers 66 per cent males and 34 per cent females). Preferential occupational structure from the studied households is given below :

Number of families	Agriculture	Agriculture and daily labour work	Agriculture and employed in different services
10	4	3	3

Eight families out of ten, have concentrated for keeping dairy animals at the new village, but per capita milk consumption is very low, *i.e.*, 0.3 litre. Only 20 per cent of families have obtained agricultural land near the new settlement area. The remaining 80 per cent of households have obtained agricultural land at different villages nearby the new location. 60 per cent of families, dislike the agricultural land given to them in rehabilitation and 70 per cent of families have obtained insufficient agricultural land compared to their original landholdings. 11.78 ha. (29.6 acres) (out of 16.12 ha. (39.36 acres) of agricultural land is for the present benefited by canal irrigation. 90 per cent of affected families have experienced non-co-operation from the nearby landholders of command area. In this group all the families are Hindu-Maratha families. No B.C. family is under this category of having medium size of landholdings. 43 per cent of families have reported about non-co-operation from the village Talathi and Gram-sevak. Because they fail to provide required certificates immediately about their land record from village abstract.

Ten per cent of families were holding very fertile land in the old location. From the new agricultural infertile land, they never obtained sufficient agricultural productions.

Yearwise affected families shifted and settled at New Chandavadi :

Year	Number of families settled at New Chandavadi
1976	9
1977	1

From this group 30 per cent households, have reported that they are neglected by the government for proper rehabilitation. 40 per cent families have cultivated trees along the boundaries of settlement plots. All of them have not benefited and obtained sufficient land as equal to that of original old village in quantity and quality. 20 per cent of households have obtained agricultural land closed to the new settlement. Nearly 70 per cent of families have obtained possession of agricultural land in time, but the remaining 30 per cent of families have wasted their minimum 4 years of waiting period to have possession of the land.

Total production of milk from dairy animals, compared with total population is calculated 0.3 litre per person and 2.8 litres to each family. 30 per cent of families have returned the loan taken from the village society already before coming to the new location.

60 per cent families are still attached to a tradition of taking dowery in marriages of their sons. This group have given humanitarian value of helping poor handicapped or sick persons for their medical help rather than investing amount in the inanimate construction of temple buildings.

The strength of educated youth is only 0.1 per cent, in which 61 per cent are S.S.C., 30 per cent are H.S.C. and 9 per cent are graduates. No post-graduate person is observed in the village.

There are 10 households who are the owner of 3.70 ha. (9.6 acres) of still available agricultural slopy grasslands at the upper part of water level of the dam and for the present they cultivate it under jowar and ragi crops.

The dairy animals have been reduced by 54 per cent from that of original dairy animals of Chandavadi.

Category "C" 5 to 10 Acres of Landholders

Five selected sample families of 44 persons (19 females+25 males) are studied from this group of landholdings.

The dependent above 16 years of age group are 27 per cent (83% females and 17% males). The ratio of dependents of women are nearly four times greater than men dependents.

Dependents below age group of 16 years are 43 per cent to the total population. (42% girls+58% boys).

Earning population is 29 per cent (93% males+7% females) to the total population.

For the present, these five families are holding 13.83 ha. (34.10 acres) of agricultural land in which 11.72 ha. (29 acres) are cultivated under irrigated crops and 2.12 ha. (5.10 acres) are under non-irrigated crops. They were holding at the original location 17.77 ha. (44 acres) of agricultural land. The ratio of obtained land is less than that of original holdings. At the original location they were producing from their agricultural land turmeric, sugarcane, rice and jowar. Now the same cropping pattern excluding turmeric is continued in the new area. Present per capita land ratio is calculated up to 0.07 ha. (7 Gunthas), and to each family it comes to 5.8 acres.

Villages from which agricultural land is obtained in rehabilitation is shown in the following Table:

Village	Khadaki of New Chandavadi	Khadaki & Panchwad	Udatare	Udatare & Saratale	Khadaki and Pachwad & Mardhe
Number of families studied (5)	1	1	1	1	1

The persons employed elsewhere are 11 per cent and majority of them are working at Bombay Cotton Mills. Per capita their monthly income is Rs. 470.

Ratio of cultivators is only 11 per cent to the total population of this group. In this category, nobody is observed working as labourer on daily wages.

Preferential occupational structure of this category is shown below :

Occupation	Agriculture and Labour work	Agriculture employed in services	Agriculture and animal rearing	Agriculture animal rearing, and employed in services.
Number of families engaged in the occupation	1	2	1	1

Out of the total, 60 per cent of families are rearing animals as a subsidiary occupation.

The daily ratio of per capita consumption of milk is 0.4 litre, which is calculated 3.6 litres to each family.

40 per cent of families have obtained agricultural land as per their choice. At present, out of 13.83 ha. (34.10 acres) of land, only 11.82 ha. (29.10 acres) are benefited by Dhom Left Bank Canal irrigation. 60 per cent of families have observed non-co-operation in agricultural work by the neighbouring landholders.

40 per cent households have explained about non-co-operation in the rehabilitation and agricultural work from village Talathi and Gramsevak. One family member of this category have admitted that he has given some more amount to the village Talathi as "bribe".

All the big landholders are powerful in the political and social fields of the village. They have taken benefits of selecting proper fertile agricultural land in the new area.

All the families of this category have resettled at new location in 1976.

Compared with other categories, big landholders have cultivated more trees in the new location of the village, within their residential plots.

All have explained, that, given assurances have been not fulfilled by the government while rehabilitating them in the new area. Only 20 per cent of families could not able to maintain their family budgets. The educated persons are 0.04 per cent.

60 per cent of landholders have obtained possession of land in time but they could not cultivate the land due to unavailability of canal water, as the work of canal irrigation is under progress. 40 per cent of landholders have wasted their 2 to 4 years "waiting period" for obtaining possession of their property.

The present dairy animal have been reduced by 50 per cent to that of old village, because of insufficient availability of grass land in the Khadki area.

Village Chandavadi Study of all the Selected Families

For the village, there are 27 per cent (19% males+81% females) of dependents above 16 years of age group. The dependent women are four times greater than that of men.

The dependents of below 16 years of age group are 40 per cent (47% boys+53% girls) to the total population of studied families.

Earning population is 33 per cent (15% females+85% males) to the total. It includes 65 per cent of literate and 35 per cent illiterate persons.

38 Sample families from the rehabilitated village, covers 49.57 ha. (112.28 acres) of agricultural land, in which there are 27.10 ha. (67.6 acres) under irrigated and 22.47 ha. (55.22 acres) under non-irrigated crops. Per capita land ratio comes to 0.18 ha. (17.8 Gunthas) and to each family 1.19 ha. (2.39 acres).

Earning population is 12 per cent to the total studied population and they are males. They are working at different services. Their monthly per capita income is upto Rs. 520.

The studied households were holding 99.51 ha. (245.39 acres) of agricultural land and now they have obtained in rehabilitation, only 48.15 ha. (119 acres) of agricultural land. The present agricultural land is diminished nearly one-half to that of original. There are 15 per cent of cultivators (10% are women+90% men). The strength of labour workers is 8 per cent (44% women+56% men) to the total studied population of the village. Nearly 74 per cent of households are rearing animals (sheep, cows and buffaloes). From dairy animals milk is obtained. Per capita daily consumption of obtained milk is 0.3 litre and to per family it is 1.8 litres.

18 per cent of families have obtained agricultural land closed to the new settlement. 24 per cent families have resettled properly. Out of 48.15 ha. (119 acres) only 38.45 ha. (95 acres) are for the present benefited by canal irrigation. 29 per cent of families are observed a good co-operation in the new area, from the agricultural farmers of their neighbouring farms. 37 per cent of families of the village have experienced that village Talathi and Gramsevak are not helping them about their land problems. Village Talathi and Gramsevak are the responsible authorities of the village, maintaining all types of village record. They are blood vanes of the farmers.

Sometimes, village Talathi and Gramsevak, becomes puppet of rich landholders and political leaders. To play games with the poor villagers about their land and to disturb family life of poor villagers is an inhuman work done by some of the rich and selfish landholders.

Only 5 per cent of the families have that opinion as they are benefited in rehabilitation.

As per their requirements 87 per cent of households have obtained sufficient agricultural land after waiting 3 to 4 years period for possessing their land.

TABLE 7.3 : **Yearwise Number of Families Settled at New Chandavadi**

Year	No. of families Settled in new location	
	Number of studied families	Percentage of studied households with total households of Chandavadi
1975	2	5
1976	30	80
1977	4	10
1978	1	2.5
1979	1	2.5
Total	38	100

TABLE 7.4 : **Agricultural Land Obtained to New Chandavadi and Availability of Canal Irrigation**

Number of families studied (38)	Year in which the agricultural land obtained in possession	Year in which canal irrigation is available to the agricultural land	Before benefit of canal irrigation: The number of years for which they could not yield irrigated crops
1	—	—	Still this family has not obtained agricultural land
1	1976	1982	6 years
6	1977	1978	1 year
9	1978	1979	1 year
1	1978	—	—
2	1977	1981	4 years
6	1979	1980	1 year
2	1979	1979	—
6	1980	1980	—
3	1981	1981	—
1	1981	—	—

Number of families	Year in which the families were settled in the new location
2	1975
30	1976
4	1977
1	1978
1	1979

TABLE 7.5 : **Waiting Period for Obtaining Canal Irrigation to the Affected Families**

Number of families studied (38)	Year of settlement in new location	Year from which canal irrigation facility were made available to their agricultural land	Waiting period after settlement in the new location for canal irrigation up to 1982-83
1	2	3	4
1	1975	1980	5 years
1	1975	1981	6 years
1	1976	—	7 years
10	1976	1980	4 years
2	1976	1981	5 years
6	1976	1974	2 years

TABLE 7.5—*Contd.*

1	2	3	4
2	1976	1982	6 years
2	1977	1980	3 years
1	1977	1979	2 years
1	1977	1981	4 years
1	1978	1978	—
11	1976	1980	4 years
1	1979	1981	2 years

TABLE 7.6 : **Waiting Period for Obtaining Possession of Agricultural Land**

Number of families (38)	The number of unutilised years by the rehabilitators waiting for possession of the land and canal irrigation to their farms (Non-productive period from their farms) (In Years)
8	2
12	3
12	4
3	5
3	6

There are 2.7 per cent of households who have resettled properly, and they are all weaker section of economic groups, having up to 1.21 ha. (3 acres) of landholdings. It should be noted that, in the new location, these rehabilitated families have improved their economy by not only the support of agriculture but also by doing on daily wages labour work.

Help from the government could not reach in time to the affected persons. Possession of agricultural land is also a time consuming process, experienced by all of them. Families of weaker sector, (small landholders) have invested more than that of obtained amount sanctioned to them in rehabilitation by visiting more than 3 to 5 times empty errants to the related offices.

There are 24 per cent of households (3% of persons to the total population), who were participated in the procession of "Morcha" for the demands of affected persons to the rehabilitation office in the year 1977.

There are 5.5 per cent of families which have been disturbed of their family life due to improper planning and utilization of obtained amount in rehabilitation.

From the village (Chandavadi), 40 per cent of families have observed that their relatives are also dam affected persons, due to their past contact within the nearby villages, which are also lost their land under Dhom reservoir.

TABLE 7.7 : **Facilities at Old and New Locations of Vyahali and Chandavadi**

Facility	Original Chandavadi	New Chandavadi	Original Vyahali	New Vyahali
1. Market Centre and Market day	(*i*) Wai : on Monday (8 km) (*ii*) Velang on Friday (2 km)	(*i*) Pachwad (6 km) on Tuesday (*ii*) Bhuinj (4 km) on Saturday	Wai : on Monday	Pachwad on Tuesday
2. Post Office	Asgaon	Jamb	Wai	Asale
3. Medical facility	Asgaon and Wai	Bhuinj Pachwad	Wai	Wai
4. Cattle Bazaar	Velang	Pachwad	Wai	Pachwad
5. Recreation	Wai	Wai	Wai	Wai
6. Ration Shop	Asgaon	New Bhivadi (Northward)	Vyahali	Vyahali
7. Cotton and Hosiery (Market)	Wai	Bhuinj	Wai	Wai
8. Bus Stop facility (But no pickup shed is available)	Chandavadi	Chandavadi	Vyahali	Vyahali
9. Library	—	—	—	—
10. Primary School	Chandavadi	Chandavadi	Vyahali	Vyahali
11. Secondary School	Wai, Asgaon	Bhuinj	Wai or Surur	Pachwad

12.	Higher Secondary School	Wai	Bhuinj	Wai	Pachwad
13.	College Education	Wai	Wai	Wai	Wai
14.	Balwadi	—	—	—	—
15.	Drinking water facility	River Wells	Tankers	River	Wells
16.	Primary Health Centre	Asgaon	Bhuinj	—	Pachwad
17.	Grampanchayat with members	Chandavadi (5)	Khadki group Gram-panchayat (2)	Group Gram Panchayat of Boriv and Vyahali (2)	Group Gram-Panchayat of Kadegaon Vyahali (2)
18.	Irrigation Facility	Electric motors on wells and River	Left Bank Canal	Wells, River	Wells, Right Bank Canal
19.	Major Crops	Rice, Jowar, Sugarcane	Sugarcane, Jowar	Jowar, Ragi	Jowar, Sugarcane
20.	Temples	(*i*) Navalai Mariai (*ii*) Vitthal Rakhumai, (*iii*) Maruti	One Temple under construction. Now all located in temporary shed	Maruti Natha, Vitthal, Shankar	One temple with all status of Maruti, Nath, Vitthal and Shankar
21.	Annual Fair	Ashwin Ashtami—Dassara	Ashwin Ashtami—Dassara	Kartika Vadya Ashtami	Kartika Vadya Ashtami.

The studied families (38) have obtained 1.37 acres of residential land from which average ratio to per family is 0.03 ha. (3.5 Gunthas).

There are 74 per cent of families are rearing animals for milk production. To 38 households, from their dairy animals, they obtained at present 76 litres of milk production daily.

Only 2.8 persons have benefited for getting employment in government services being the dam affected persons.

All the villagers including children remembered the past happy incidents of old village festivals, like annual fair Dasara, Divali, Padwa. Their mind was tossing within the vision of old village environment.

Economic life of 58 per cent of households have been disturbed and diversified. This ratio is decreasing as per increasing rate of landholdings. Social and economic family life of affected persons, becomes more stable as per increasing rate of landholdings. 34 per cent households have accepted dowery system. Instead of dowery, they formed a new word for it as "Gift". Traditional ties are very strong and too rigid that they cannot change such bad traditions for the betterment of society. In writing, many of them have explained that they are against of taking dowery in marriages. In the marriage ceremony, to give party or dinner to all the villagers, is a tradition still continued and they do not feel it as more showy and expensive type of investment. They take loans for celebrating annual fair, Satyanarayan Puja, and other functions even they are unable to maintain such type of investments, which are not worthwhile.

Being dam affected persons, settled in infertile area, persons of their relatives have formed a gap and broken closed ties with them. There is 6 per cent unemployed population within the studied households. Literate persons are 12 persons (68% S.S.C., 26% H.S.C., and only 6% graduates).

39 per cent of households of the village have wasted for obtaining possession of agricultural land—the period of 3 to 4 years.

To all of the studied families, there are at present 12.80 ha. (31.28 acres) of land available above the water level of the dam, at the original location. In it 9.10 ha. (22.22 acres) (maximum acreage), are covered by small landholders, of below 1.21 ha. (3 acres) of landholdings.

Strength of dairy animals have been reduced by 50 per cent in the new village. 15 sheep are remained out of 322, from 38 families at the new location.

Cattle was very important source of the valley for farm uses, and for local use of milk production. There were extensive grasslands, and sometimes they used government forest land for grazing purposes. Now at new

TABLE 7.8 : **Landuse of Area Obtained and Allotted to shifted Population (Vyahali, Chandavadi)**

Name of the village	Landuse			Grass land
	Group A	Group B	Group C	
Chandavadi New	Jowar, Pulses	Groundnut, Wheat	Sugarcane, Rice	Except Govt. Land of mountain slopes to the east, no grass land is available to the new village
Vyahali New	Sugarcane, Jowar	Wheat. Rice	Pulses, Turmeric, Vegetables	Except Govt. Land under mountain slopes of Vairatgad "Asale Bhavani Hill" range, no grass land is available to the village.

Group A : Major Area under cultivation
B : Medium Area
C : Small area.

locations there are no grazing lands. So majority of them do not keep cattle.

There are 2.6 per cent of households, waiting to obtain agricultural land from the government. The landless families are working in the agricultural fields of nearby villages.

In general, compensation of amount paid to the affected persons on account of acquired agricultural land for the project in the varied scale from Rs. 1,000 to Rs. 6,000 per acre. Medium and big landholders have constructed houses with bricks and roofed by tiles and tin-sheets. Poor families have constructed huts and kutcha houses.

At the new locations, the settlements are observed, which are based on old traditional methods on castes and religions.

8

The "Shift" as Perceived by the People of Concerned Social and Political Leaders and Members of the Administrative Staff

Views of Social and Political Leaders

THERE were eight social and political leaders were interviewed in relation to understand their views about rehabilitation problem of dam affected persons. As per their experience, the appointed rehabilitation committee could not work satisfactorily, because there is no proper co-ordination in between the members of the committee and the government personnel and the displaced persons. No voice of representative of dam affected villages is respected by the committee. To obtain sufficient benefit of local resources in the new environment for the affected persons "Footing Plans" should be prepared and submitted to the government by such committee. A remedy suggested by social workers, is to form unity of all dam affected persons for asking government to their basic demands and present the difficulties they have to face in rehabilitation. The demands should be social and realistic. They should not be coloured by political reasons. To check current practices on the part of government servants, there should be a social movement. Market values of assets taken over from the dam affected persons, should be equated with rehabilitation costs. Agricultural land should be given at very close to their settlements, or they should be adjusted and allowed to construct residential shelter within their agricultural farms. The house for a poor villager is always his shelter, used for multi-purpose work, i.e., kitchen room is a bed-room and in it, sheep, goats, hens are also protected.

Family-wise land distribution is a proper way for rehabilitation. Slab for acquiring agricultural land should be decided in such a way, that all the displaced families can be merged within the command area.

The social and political leaders, who are not actually dam affected persons but who are working on the "rehabilitation committee", were not able to understand the difficulties of displaced persons and they were also not able to take interest in work. Such leaders do not eager to be associated within the work of rehabilitation. It becomes difficult to take the representatives of dam affected villages, due to they are not in mood to work under unstable unsheltered position of their family. The demands are not granted sometimes due to insufficient membership of the committee and sometimes improper decisions are passed in it.

Different departments are collecting the data of village area for research purposes. Correct data, through such agencies, with its results, should be communicated to the village and for this work, an agency should be appointed by the government so that research work will not be remained on paper only, and it should be implemented quickly.

From the rural area, strength of public is utilised for obtained local demands from the government, which afterwards benefited for only their leaders. All the villagers are not experts in that way. So they should be guided through accurate plans.

Political leaders, observed too much reserved, they were not frank while expressing their views about the problem. One leader suggested hesitatingly that the scale slab for land acquisition of command area, should be lowered from 10 to 5 acres, and said that his name should not be disclosed. The reason for this is that, as he does not want to take blame from the public which is his "Voters Bank", the slab suggested by him is against the wishes of the people of canal benefited area.

Due to Dhom dam, village Asgaon is shifted at Vaduth, Tal. Satara and "President, Rehabilitation Committee of Asgaon", have explained about a resolution Number RPK/1078-555/R/1, passed on 15.11.1978, by Government. Under this resolution Government can return the acquired land from command area to the original landholder, if he proves his family is a divided family, this way the holdings get converted into those of small landholders, i.e., of below the acreage of slab fixed for land acquisition by the government. Thus many big landholders, by formally breaking their joint families. This is a chance for Talathis, circles, to take bribes from them for doing the needful of dividing and recording their family land within the family members, (as small landholder) in the village abstract.

Due to this resolution the families of Asgaon lost the land, being taken back by the original landholders of command area, providing to the court, by dividing their families as "small landholders", after obtaining agricultural production for three years.

This resolution of 15.11.1978, is now superseded by the new resolution of Government of 14.12.1982. Now under the new resolution, the provision of appeal to the Collector, or to the state government, by the landholders of command area, is dropped. A function held in October 1970, at Kukadi Project (Dist. Pune), was obstructed by Panshet landholders and the chief guest a political leader of the function said "problem of rehabilitation of dam affected families should be solved immediately. We do not know the actual problems of dam affected persons their unstable position. There position is like 'the wearer knows where the shoe pinches".

If a landholder, who is lost under submergence of dam, 16 acres of mountainous infertile land, he is not called as "dam affected person", and he is not entitled to obtain facilities of rehabilitation. There were many dam affected persons engaged in the work of construction of the dam as labourers.

One social leader, explained that there is a common answer given to dam affected persons by the member of the staff of rehabilitation or land acquisition office and it is that, "I can't do anything for you. I have completed my work. Now you can go in a court to settle your matter". Two illiterate women of Dhom village, came to L.A. Office at Wai on 30.4.1982, to take the amount of compensation of Rs. 12 to each, but somebody from the village told them that you are obtaining Rs. 1200 amount to each from the office. Due to mis-interpretation of Rs. 1200 they were in happy mood, came with their grandsons, planning for a dream to purchase sweetmeats and clothes. Actually they obtained only Rs. 12 each, became nervous and returned to their homes with their crying grandsons and no sweetmeats, no clothes.

One social worker explained about the position of Koyana rehabilitation. The displaced persons of Koyana dam are very poor. As per rules of 1965, by the state government, they obtained compensation of their agricultural land, as per rate for infertile land per acre Rs. 60 and for fertile land Rs. 300 only. For the present (March 1982), there are still 132 unrehabilitated families of Koyana project from the following villages Ahire, Tapole, Bamnoli, Pali, Valane, Avalane. There is a dense forest in the upper part of Koyana valley, Foot-paths are common in the region. People are very backward and traditional. They do not know how to get contact with related offices.

Rehabilitated Village

Gogave is an ideal example which is situated in infertile area. The land is not suitable for agriculture. From this village Shri Kadam have made a successful experiment by cultivating 20 acres of infertile land, by the following trees: (i) Nilgiri, (ii) Suru, (iii) Bamboo. There are 65 members in his family. He cultivated (in the period of 1965 to 1968), 20,000 Nilgiri plants obtained the income of Rs. 6000 per acre.

Past president (Sabhapati) of Jaoli Panchayat Samittee, explained a problem of rehabilitation of Kanher Project. In the year 1981, some villages were shifted to Pandharpur (Dist. Solapur) and again many of them are called back in Satara District. In this way, the affected families were shifted more than twice of their settlement location at different places. They are now very tired for changing of village location (i.e., Villages: Kaleshi, Venegaon and Targaon). Mr. Sabhapati, gave an example, that person related to a political leader has obtained a good location of 4 acres of agricultural land at Vadhe, Tal. Satara immediately though he was partly affected by Kanher Project.

Amount obtained for construction of a house is very small and they could not even built one-fifth part of their buildings.

Views of Members of Administrative Personnels Connected with Rehabilitation

Twenty-five members of administrative personnel were interviewed. They explained that, it is difficult task to the servants of irrigation rehabilitation and L.A. departments for implementing government rules and regulations for settling the matters. They have to face many difficulties for acquiring the land from command and catchment area of the project, shifting of affected villagers at new locations, and distributing the land to displaced families. This is a challenging and time consuming social work.

For the present, rehabilitation work is done by 4 to 5 different departments of State Government with many other agencies like Z.P., M.S.E.B., Land Record Office, Land Acquisition Office, Irrigation Department, etc. All these offices are located and working from different places. It is essential to have proper co-ordination between all these offices. The papers and files are moved from one to the other office, consumed too much time and late decisions obstruct the work. Internal division of work is required for smooth and strict administration.

If the L.A. Office has obtained a very fertile land from command area, it becomes difficult for land acquisition office, to select the families due to limited available land and unlimited demand for it from the displaced families.

Everybody tries to get residential plot closed to the main road in the new village. Distribution of plots is again a headache to the government offices.

Bad practices are observed from some affected and benefited landholders. Many joint families are converted into divided families in the village record on paper to obtain profit by rehabilitation. Due to this, benefited landholders are saved their land from L.A. Office and affected landholder cannot obtain more agricultural land and residential plots at new location from command area.

Due to obstructions raised by the landholders of command area, work of the surveying land, becomes a very slow process. From command area, the landholders are trying to see that the canal route should be constructed from outside of their agricultural farms. At the time of survey the land for demarcating boundaries of their farms, they remained absent and raised objection against the L.A.O. of about the incorrect measurement of the land.

It is found that a dam affected family has taken a plot at new location of Asagaon, at the same time this family have obtained another plot at rehabilitated Vyahali. With one concession, this family have taken two residential plots at two locations and misguided the rehabilitation office. To control over such practices, co-ordination between irrigation, Land Acquisition and rehabilitation office, with related Talathi of the village is essential.

The administrators explained that land acquisition of catchment area of Dhom Dam was very easy than the work of L.A. from command area. All the related administrators were agree to this opinion that it becomes impossible to provide all the basic facilities in time to all the dam affected villages.

At the time of selecting new site for village settlement, it should be considered of providing facility of easy water supply for drinking purpose. The area should have at least the source of well irrigation facility. Village Asgaon is located near Vaduj, Tal. Khatav, Dist. Satara, where there is no provision for source of water supply. For such villages, government is agreed to invest Rs. 1 lack to the work of providing drinking water through 'pipeline water supply scheme'.

At the time of acquiring land by the government there is a notice period of 51 weeks which should be reduced up to 2 weeks only.

For Dhom Project, sufficient and fertile agricultural land is obtained from the command area villages of Limb and Mardhe, because, they are saved from shifting business emerging by "Patkhal Borkhal Dam", which is cancelled by the government of Maharashtra, as per demands of the villagers of Limb and Mardhe.

Under the slab fixed by the government there is also a provision to acquire 20.23 ha. (50 acres) of land from a village of command area. According to the act formed and came into force from 1973, the slab is revised up to 2.83 ha. (7 acres) to eight family members. Now the slab should be again revised up to 2.02 ha. (5 acres). Reconsolidation should not be repeated henceforth.

The rules of rehabilitation to fully and partly affected villages are different. From a partly affected village if agricultural land (excluding residential land) is acquired for the project, the village is only liable to obtain only agricultural land. No residential plots are given to them. For special cases, there is a provision to provide agricultural land and residential plots at new location.

It is observed that, no calculation has been made about the loss of vegetation under submergence of Dhom Dam. Under Kanher Dam of Satara district 2000 fruit trees are lost and the loss is calculated to Rs. one lakh.

The affected villagers are more co-operative and helpful than the villagers from command area. The villagers of command area were observed strictly against giving their land to the dam affected families. Before rainy season the work of shifting villages have been completed by the government.

Henceforth, in rehabilitation the complicated cases should be tackled by a special officer by visiting the actual fields of problems in the related villages, for taking quick decisions. A vast period is consumed by court matters for taking decisions of rehabilitation problems. A separate court (which should be mobile) in the limitation of command area may be appointed as per necessity of the work.

Sometimes record of land in the village abstract of 7-12 is not found correct due to illiterate landholders.

Members of the staff of rehabilitation office explained that in rehabilitation 50 per cent of the families have obtained infertile and insufficient land.

Rehabilitation is not profitable for small and medium landholders who have obtained agricultural land away from their residential area (at a distance of 4 to 8 kms.). There is a rule to provide agricultural land to the affected families within the distance of 8 kms. from the new village with the help of Executive Engineers of the project. The engineers always try to select a land on either sides of the canal, or benefited by canal irrigation.

The remained agricultural land at old sites above the water level at Dhom Dam, should be acquired by the government distributed to the partly affected villages residing at protected part of their old sites. The work of rehabilitation is distressive and discriminative, feels to the consciousness to the government employees, because they could not satisfy dam affected persons due to their limitations of work.

The dam affected persons, coming with their applications for obtaining particular facility or representing the problems and complaints, do not know to which office or offices they have to approach. They started to blame the government offices, that nobody from the offices, giving full co-operation and helping to them about their matters sympathetically.

To complete a work and to get a contact with proper office, they have to spent lot of money and lot of time by visiting more than twice to four to five related offices at different places. It is found that, one Talathi is working for three to four villages at a time. His work should be limited by fixing a particular quantity of landholders, so that he can maintain correct agricultural landuse of the landholders. The quality, ability and qualification of Talathi and Gramsevak should be improved. There is an example that a family of the village Limb, Tal. Satara has lost agricultural land for canal construction, but the village Talathi, is still (1982-83) taking revenue charges of the land. Sometimes late declaration of award of acquired land, the farmers have to bare unnecessary burden of paying revenue charges of the land.

Information obtained from the rehabilitation office Satara :

	Name of the village (studied)	No. of unrehabilitated households	Partly resettled households
1	Vyahali	3	4
2	Chandavadi	2	5

Reactions and Problems of Dam Affected Persons

It is also observed that, village Talathi, is not taking immediate actions, of keeping up-to-date record of the village and making the changes suggested by L.A. Office. At the very late time, Talathi always expects that the landholder, should come to meet him and request him to do the necessary changes made by the government.

The major complaints and problems of dam affected families are the following :

(*i*) Fixation of price of acquired land, trees, buildings, and other properties.

(*ii*) Distribution of loan as per quality of land to the landholders.

(*iii*) Distribution of land and residential plots. (Land for land is the best policy).

(*iv*) To form good relation in between rehabilitated villages with nearby farmers of command area villages.

(*v*) New villages should be located near the growth centres of the region and social and other facilities should be provided to it.

(*vi*) To provide employment facilities and occupations for utilization of manpower to the betterment of society and the region.

(*vii*) To settle the court matters in rehabilitation work immediately.

Their major reactions were as follows :

(*i*) They were eager to form a unity of displaced persons for asking to the government to solve their present social and other problems.

(*ii*) They were not against any government servant or any political leaders but they were asking for proper guidance and help from them. Displaced persons should not be neglected by any Government servants. Co-operation and good relations should be developed in between the leaders, officers and uprooted unsheltered persons by Dhom Project.

9

A Brief Survey of Some Villages for Evaluating the Benefits from the Dhom Project

Introduction

SIX villages with selected 22 landholders from command area of Krishna Dhom Project were studied. All of them have started for giving their part of land to dam affected families. A sample of the villages are (*i*) Pachavad, (*ii*) Chindavali, (*iii*) Anewadi, (*iv*) Pasarani and (*v*) Ozarde of Wai Tahsil and (*vi*) Village Mardhe of Satara Tahsil.

All the families, except one, have benefited by the cheap source of canal irrigation, and have improved their economy due to irrigated crops, and now they are economically stable and can be very easily maintain their family budgets.

Increased Area Under Irrigated Crops

The above villages were well known for dry farming crops, *i.e.*, jowar, hybrid, bajra, pulses and groundnut crops. Well irrigation facility was dominant and the agricultural fields at the vicinity of well, were important for irrigated crops like sugarcane, turmeric and vegetables. The area under well irrigated crops was very limited. Now due to canal irrigation facility, the area cultivated under irrigated crops is increased and till to the present it is extending along the sides of canal area.

From the study of 22 sample families of the above mentioned villages, the major observations are obtained, which are as follows.

The studied 22 households, with 158 population (90 males+68 females), are holding 130.9 acres (52 ha. approx.) of agricultural land in which 27.20 ha. (67.10 acres) are under irrigated and 25 46 ha. (62.39 acres) are under non-irrigated crops. Out of total irrigated land there was 6.70 ha. (16.24 acres) land under well irrigation which is now converted into canal irrigation.

Electrified wells were providing water to agricultural crops. Due to canal irrigation facility the electric pump sets are remained unutilised.

Per capita well irrigated land was only 1.4 Gunthas. Now through canal irrigation facility the irrigated land is increased by 0.16 ha. (16.5 Gunthas).

Four big householders out of 22 have given 7.27 ha. (17.39 acres) of agricultural land to the dam affected families. And for 7.27 ha. (17.39 acres) the obtained amount of Rs. 55,700 (as a total price of the land) as compensation. The rate comes to per acre Rs. 3,100 approximately. Land acquired by the government under canal route is 1.42 ha. (3.21 acres), and for that the landholder have obtained amount of Rs. 14,275 as a price of the land. The rate comes to per acre Rs. 4,447.

The villages of command area located at the eastern part of Wai Taluka along the Krishna River Course, were well known for lifting up the water from the river through oil engines and by well irrigation facility. Now, after availability of canal irrigation facility within the same area, no change is observed within the agricultural landuse and the same cropping pattern is continued (*i.e.*, Bhuinj, Pachwad and Ozarde villages). The area which is located away from the course of river Krishna was under dry

TABLE 9.1 : **Cropping Pattern of Command Area Villages**

Crops	Original cropping pattern before availability of canal irrigation facility cultivated area (in acres)	Present cropping pattern after availability of canal irrigation cultivated area (in acres)
Jowar	34.00	12.24
Wheat	2.20	6.20
Groundnut	28.30	9.11
Rice	4.30	4.34
Pulses	11.00	8.00
Bajra	0.20	—
Turmeric	3.20	8.00
Sugarcane	10.20	31.23
Ginger	1.20	4.00
Total	97.00	84.32

crops (jowar was dominant crop). It comes now under canal irrigation, *i.e.*, irrigated farming.

The canal irrigated land now completely converted under sugarcane cultivation followed by turmeric and ginger, which was mainly producing jowar and groundnut crops. This might be an impact of canal irrigation and due to nearness of sugar factory, at Bhuinj Kisan Veer Nagar.

Cropping pattern of selected 22 households from command area of selected six villages is given in Table 9.1.

The land revenue charges paid by the studied landholders were Rs. 726.50, before canal irrigation, and for the present at the charges are increased up to Rs. 920. The income from the agricultural productions is increased nearly four to six times greater than that of original.

Only thirty per cent of canal benefited landholders are agree to help to the dam affected persons if any amount is collected as an aid by proper agencies.

A co-operative lift irrigation scheme which is formed by 17 members of the village Anawadi, Tal. Wai, constructed on Dhom Left Bank Canal. Now the scheme becomes fruitful to irrigate for 5 per cent of village agricultural land. There is restriction on sugarcane cultivation under canal irrigation within the village area. Hence they have cultivated cash crop turmeric instead of sugarcane.

The total annual income, before availability of canal facility, from 22 studied households, was Rs. 44,000, which at present after canal irrigation is increased up to Rs. 85,000.

Per head annual income was Rs. 278, which is now calculated as Rs. 537.

Fifty per cent of families of medium landholders are now under the stable economic group, but lack of proper planning, skill, and lack of knowledge of local resources, they are remained still unprogressive and backward.

Animal rearing for dairy production, is also prosperous occupation to this area. The villagers are inclined to have she-buffaloes rather than the cows. The people do not have proper knowledge and importance of cow for milk production. Cow milk contains more vitamins, and it is more nourishing than the milk of buffalo.

Capacity of women for physical hard work in the landless families is greater than the women belonging to the families of big landholders.

Most of the families of B.C. communities are coming under small landholder category. B.C. people who are now Buddhist, performing religious songs of Buddha Dharm at the time Buddha Jayanti or at the time of marriage ceremony, but nobody knows the meaning of such prayer of "Buddham Saranam Gachhami. . . .".

TABLE 9.2 : **Command Area Village—Land given to the Affected Families, Canal Benefited Area and their Annual Income from Agriculture before and after Canal Irrigation**

	Name of village						
	Pachwad (Tal. Wai)	Chindavali (Tal. Wai)	Anawadi (Tal. Wai)	Pasarni (Tal. Wai)	Ozarde (Tal. Wai)	Mardhe (Tal. Satara)	Total
Sample families studied	6	6	2	2	4	2	22
Population: Male	31	20	6	9	17	7	90
Female	25	19	5	8	7	4	68
Total	56	39	11	17	24	11	158
Agricultural Land							
Irrigated (in acres)	23	14.10	1.0	3.0	23.0	3.0	67.10
Unirrigated (in acres)	19	6.35	7.0	9.0	10.21	10.23	62.39
Total (in acres)	42	21.5	8.0	12.0	33.21	13.23	130.9
Land given to Rehabilitators (in acres)	—	10.0	—	—	4.39	3.0	17.39
Land Lost under canal route (in acres)	1.6	00.8	0.20	1.15	00.2	0.8	3.21
Amount obtained as price of the land for canal route	6,300	500	4,000	2,100	375	1,000	14,275
Amount obtained as a price of Agricultural land given to the rehabilitators (Rs. in thousand)	—	30	—	—	20	6	56
Canal benefited area (in acres)	21	13.10	1.0	2.0	18.21	5.23	61.14
After canal irrigation per head Annual income in Rs.	356	860	359	470	608	454	538
Before canal irrigation per head annual income in Rs.	229	237	318	411	370	218	275
After Canal irrigation income in Rs.	19,900	33,500	3,950	8,000	14,600	5,000	84,950
Before Canal irrigation Annual income in Rs.	12,800	9,250	3,500	7,000	8,900	2,400	43,850

The movement seems to be again traditional. They accept anything which they feel as the movement of prosperity.[1] No doubt, that they accept new trends in the society, but the trend becomes unsuccessful and traditional. This position is not helpful to give mental peace and progressive attitudes to the society.

Due to faulty planning, employment of agricultural part has been grown up to unmanageable levels.[2]

It is sure that due to canal irrigation facility, if planned scientific agriculture is developed in the villages, the economy of the area will support to agricultural labourers by providing agro-based small scale industries. Even it is necessary to provide, the knowledge to the farmers, for careful utilization of water and land by deciding particular cropping pattern suitable to the area.

MFAL and SFDA agencies should give a proper support to all the big and small landholders for generating rapid change of the economy of rural area.

Impact of Rehabilitation of Dhom Project over the Market Centres

Wai and Panchagani were the market centres for all the dam affected villages. The villages were important for rice production. Now, due to loss of vast rice cultivable land under Dhom Project, the production of rice is decreased. Wai market for the present is important for vegetables, turmeric, sugar, groundnut and jowar. For Panchagani, the area was important to provide cheap labourers to work in bungalows and in hotels. "Pachwad" (Weekly Bazar on Tuesday) and "Wai", are the market centres to "New Vyahali". For Chandavadi, "Bhuinj" (weekly Bazar on Saturday) is the nearest market centre.

The trade centres: Bhuinj and Pachawad are located on "Pune-Bangalore Highway" and Dhom, Chandavadi, Renawale, Wasole, Bhivadi, Khavalim, Govedigar, are rehabilitated near to this highway. (Fig. 4.2). At Pachwad, in summer season, temporary open air theatre in a big tent is opened for cinema entertainment. Most of the farmers of this area are free from the agricultural work, in April and May. This is the relaxation period to them and many villages every year celebrate their annual fair—Jatra in summer.

Name of the market centre	Major facilities available
Wai (Taluka Place)	Medical, Educational (College level). Shopping, Hoteling, Banking, Recreational facilities (All the administrative offices at Taluka level).
Bhuinj	Sugar factory, medical, Educational up to H.S.C. Shopping, Hoteling and Banking.
Pachwad (Annual Bazar for 10 days in summer)	-do-

At Bhuinj and Pachawad, "electric motor repairing shops" are in difficult position as nobody from the area now depends on electrified wells due to available source of canals.

Many of the Chandavadi villagers are always used to visit Bhuinj to purchase their different requirements and needs, i.e., to purchase cloths to take medical help and for many other things.

Cultivators sale their agricultural goods, *i.e.*, jowar, wheat, groundnut, cereals at the market centres of Wai, Pachwad and Bhuinj including vegetables, sugarcane, turmeric, ginger, are the main cash crops of canal irrigated area. The region will be very prosperous in the future, if the landholders will implement most modern scientific agricultural methods within their farms and if they will properly utilise develop and conserve their local natural resources by systematic planning.

One has to bare the loss, while other has to gain the profit. This is the position of displaced persons by Dhom Dam with benefited villagers from command area. They lost permanently their original environment and their agricultural property. But due to their shifting, government have constructed Dhom Dam which now benefited to the villagers of command area by canal irrigation facility. For dam affected villagers, the only profit, is that they acquired their village locations at the developed and more advanced eastern part of Wai Taluka.

Wai, Pachwad and Bhuinj are slowly supporting to establish different types of occupations of the area. In the near future, Wai, Pachwad, and Bhuinj will be important growth centres and urban zone of eastern part of Wai Taluka.

As population is increasing in the area no market centres are adversely affected due to shifting of dam affected population.

REFERENCES

1. A few Geographical Sample studies of rural landscapes in Wai Tal. (Page 53) Research Project of Shivaji University, Kolhapur, Shri Prakash Sawant, 1982.
2. Sen, A.K., 1973, India—Five-Year Plans (Mimeographed).

10

Possible Guideline for Rehabilitation of Displaced Families by Irrigation Project

FORMULATION of project is a complex process. It should cover the following three stages :

First Stage

(*i*) Plan for irrigation project: showing area under submergence of a dam and showing a map of benefited part of the project with canal routes.

(*ii*) Plan for acquiring suitable land of command area: for merging dam affected families and plan for land acquisition of the catchment area.

(*iii*) Plan for work with strict and quick administration, time limit and total expected budget for the work. Responsibility to shift and resettle a village should be with only one office which is final authority for all types of proper decisions.

Second Stage

(*i*) Shifting of dam affected villages, resettling them and providing them all the necessary facilities, should be started at the same rate of construction work of Dam and work of canal and sub-canals constructions up to the agricultural farms.

(*ii*) Provision and plan should be worked out actually by merging particular families in particular villages of command area. The value of acquired land, residential buildings and other properties can be approximately calculated, with the affected families and four times of such calculated amount should be used by the government for rehabilitation work. (Excluding expenditure of construction work of the dam and canals).

(*iii*) As per landholdings, compensation of land should be given.

(*iv*) After location at new settlement, for the period of 4 to 5 years, the affected families should be provided in the region small scale industries and any other work for them as a source of income.

(*v*) Social institutions, big landholders and rich persons of command area and big industries, should be informed to help the social work of "Rehabilitation".

Third Stage

(*i*) Provision of water management and fixation of water rates according to cropping patterns. The water rate for cash crops should be more than the foodgrains.

(*ii*) Distribution of water is a responsibility of Canal Supervisor, but there should be "Surprise Visit Officer", who can control strictly on him, for misuse of water distribution to the landholders.

(*iii*) Growth of Economy of benefited villages should be recorded, due to canal irrigation agriculture.

Before starting the construction work, it is found that big landholder of command area are trying to purchase the canal benefited agricultural land in cheap rates, as a non-irrigated land.

To find out equivalence of non-irrigated land with irrigated land from catchment area, and to provide the affected families, canal irrigated land from command area the following formula is suitable and can be used for calculations. It is true that the agricultural production per hectare from irrigated land is generally five times greater than that of non-irrigated land. Due to irrigation project "Canal irrigated land" of command area is in this way becomes more productive than before, which resulted improved agricultural economy of the people.

The actual land acquired—under submergence of Dhom Dam is 2308 ha. (5704 acres), included in it, settlement land, (culturable waste), forest, grass lands, barren lands and cultivated land from all the partly and fully affected villages.

After excluding rockey, barren and mountainous grass land area—and forest area by 60 per cent to the total of 5704 acres (1971 census), the remaining 40 per cent of the land (2281 acres) is required to the government for Land Acquisition purpose from the command area.

From command area of Krishna Dhom Project, the total canal

irrigated land is 63200 acres. Out of this area, we have to provide to the dam affected villages only 1/27 of the area.

From Left Bank Canal : 19160 ha. (47900 acres) Irrigated land.
From Right Bank Canal : 6192 ha. (15300 acres) Irrigated land.
Total 25352 ha. (63200 acres) Irrigated land.

Due to left and right bank canal of Dhom Project, the total 72 villages have been benefited from the Wai, Jaoli, Satara, Koregaon talukas of Satara District. If on an average from the command area, there are 25 big landholders, having more than 5 acres of landholdings, and from each of them 2 acres of canal irrigated land is obtainable, which comes to 50 acres from each village. Though, 75 villages are benefited by canal irrigation under Dhom Project, we select only 50 villages by which we can obtain 2500 acres of agricultural land. Or if we select 1500 big landholders from command area, with the rate of 2 acres of acquiring the land 3000 acres of land is obtainable in this method. Our necessity of land for Dhom rehabilitation is 2281 acres, which is less than that of the actual obtainable land.

In the command area there is also available mountain slopes, which are infertile, under grass lands, can be provided to the different settlements at new locations. Drinking water can be provided them through pipeline water scheme from wells located to the lower part of the region.

The above example is only taken to find out a solution over "rehabilitation" problem, if properly planned to merge the dam affected families in command area.

To check and control over the process of dividing big landholder of joint families for showing as "below slab landholders", medium landholders should be given more importance for "land acquisition process" as they are larger in numbers. From them, slab should be fixed in such a way that from each family little quality of collected land will be available from command area and sufficient agricultural land will be with the government for rehabilitation purposes.

For planning of resettled village, it should be remembered that a village is not formed only by a group of persons. The cultural and social community life should be exposed by utilising constructive co-operative human relationship within the resources.

All over the area, size of the agricultural farms are very small. Land consolidation is failed due to separation of joint family system, occurred due to growth of population in rural families. Marginal and medium landholders are un-productive landholders due to poverty and for them population becomes pressure over the land. Such land should be developed through scientific ways by suitable cropping pattern. Commercialization of agriculture is necessary. Even grasslands and infertile land can be used for cultivation of sisal. Forestry, grazing and other many suitable occupations for the environment of the area, should be introduced. The present 75 per cent of population is busy with agriculture and primary type of occupations. From this, atleast 1/2 population should be removed from agriculture into agrobased occupations and for resource development work in the area. Rural man-power planning is necessary. Population and resources should be compared and measured by qualitative (but not quantitative) measures for changing the "Krishival" into "Scientific and agro-based occupational" new village structure.

11

General Observations and Conclusions

The Lessons Emerging from the Study

RESULT of Irrigation Project: Villages shifted in the year 1971 and 1972 are somehow obtained better rehabilitation facilities than those who have shifted afterwards.

Formation of two different economic pictures: Here are two different economic pictures observed from catchment and command area of the project :

(*i*) Unstable life of dam affected villagers, they are anxious about their future. Their family life is scattered over different new locations. Majority of small and medium landholders are observed uneasy due to their deserted life.

(*ii*) The happy life of the villagers from command area as they are benefited by Dhom canal irrigation facility. Irrigated land of the area is increased and it results high yield from agriculture. They are stable in economy.

THE SHIFT TO THE NEW SITES IS NOT EASY

Emotional Problem

It is emotional and environmental impact over man to leave the old home with old environment, in which he has spent and accustomed much of his life time. For shifted families, it becomes very difficult in adjustment to the new home given to them, i.e., "the sense of belonging". It is suggested that the government should provide them in time, at least their basic needs and resources or fertile land, so that they can rebuilt the new environment as equal to that of old environment. This is possible not by formation of small new settlements or villages but by merging affected families within all the benefited command area villages.

ADMINISTRATIVE PROBLEM

Lands Acquisition

Cultivable land for compensation is not obtained at a place and in time in the command area. For Land Acquisition Office, to acquire the land for rehabilitation of any village, is a very difficult work. This is because in command area, to save landed property from Land Acquisition Office, and in the catchment area in order to obtain maximum benefit from the government, the big landholders divide their joint family and convert themselves under "small landholders" on record only. This bad practice should be controlled by strict rules and regulations or by legal provisions. Land distributed to the affected families from command area is mostly infertile, because no fertile land was available to the government as per slab, and the "selection of land for acquisition purpose" is laying with the landholder of command area, who always prefer to give to "L.A. Office", his infertile land. This selection of land for acquisition must be done by the L.A. Officer. Well irrigated and productive land should be acquired by the L.A.O. for this purpose, to obtain sufficient and productive land to rehabilitation work, the slab for acquiring agricultural land from command area, should be revised from 7 acres (1973) to 5 acres for 8 family members.

Possession of Land

From the command area, original landholders do not give easy possession of the surrendered land to the affected families. Again a provision should be made in rehabilitation act, so that, possession of land will be quick and easy process.

Lengthy Process of L.A.

The present process of acquirement of agricultural and settlement is lengthy. The responsibility of taking final decisions of any matter should be handed over to one rehabilitation or L.A. Officer, for a particular area. There should be one Land Acquisition Officer (L.A.O.), who is also responsible to acquire the land from catchment and command area. He should be given particular area or villages to shift from the catchment area and to resettle them in the command area. This work should be completed on time and by the same office, so that, for landholders, it will become very easy to get contact with one office only for all the related matters. Human elements should be given equal and more importance with nature for rehabilitation work. To solve the personal difficulties about rehabilitation matter, a person if he does not know at where and to whom he has to make contact or to visit, his ample time is killed for finding out proper offices for completion of his work. Even offices related to rehabilitation are located at different offices is on a great distances from each other.

Machinery working for Rehabilitation

The machinery working for rehabilitation is so large and so complicated that there is no proper co-ordination within them. Four to five different departments are working for rehabilitation of dam affected families, but no proper control is observed within all the related offices. At present the complete expenditure of rehabilitation facilities are provided and incurred by the "Irrigation Project". The "Irrigation Project", provides and sanction the amount to Zilla Parishad (for development of road constructions and other development facilities to the new village), to M.S.E.B. (for provision of electricity) and many such dapartments related to development work. It becomes very difficult and complicated to provide the amount and to collect the accounts from them at an early date. It is suggested that all the responsibility of the work should be handed over and given to the "Irrigation Project" so that the project as one of the main responsible unit, can form a good co-ordination of work for immediate better results.

For the present, the work of distribution of agricultural land is done with the help of rehabilitation committee which is finally accepted by the D.R.O. (District Rehabilitation Officer). Political and other prominent members are included in the committee, though the members are not properly well-versed in this work. It is also found that the personal interests of local leaders has formed unequal distribution of agricultural land and they always tend to give profit to such affected persons who can help to them to add their votes at the time of election in "Voters Bank". Rich farmers have managed to obtain sufficient compensation. Poor families are remained backward and hungry in the queue of waiting period for next unanswerable period. As per the opinion of Mr. Zimmerman, it should be kept in mind that a farm community is a poor melting pot. Heterogeneous people and culture do not mix well in such a community. Many attempts to establish this kind of community have failed miserably and have often ended in profound trouble and strike".[1]

It was a practice of preparing a map of showing the command area and the ways of canal bed by the office without visiting the actual fields of the area. This is resulted by some mistakes of showing unirrigated land under canal irrigation. It is always remembered that map of command area should be prepared by surveying the actual fields so that no mistake should take place for measuring the actual canal irrigated land. This is a responsibility of rehabilitation office or the project to produce accurate map of canal benefited area with the routes of canals. This work should be done by taking help from the Executive Engineers of the project. No approximate area is calculated and shown on maps of command area. Once the alignment is fixed, then the process of land acquisition will become very easy.

In Maharashtra, the Irrigation Department has helped to all the affected families by providing a free of charge facility of transporting residential and other important materials from old to new village area.

Shifting and Resettlement Work

Two to four villages should be shifted and resettled by one administrative officer. There would be more staff appointed as per requirement of the work. To each officer a jeep should be provided to make easy and quick contacts with the landholders of command and catchment area. To save the time and money, it is already suggested that the work of land acquisition and land distribution should be done through one office. For example to shift the villages Vyahali and Chandavadi from their old location by acquiring their agricultural and settlement land for "Dhom Project" and to settle them in the command area for Vyahali at Kadegaon area and for Chandavadi at Khadaki village area are the duties should be done by one office. This is essential because documents and certificates related to land are not obtained by the affected persons in time.

Payment of Compensation and Amount of Loan

Payment of compensation and amount of loan sanctioned for construction of residential shelter, should be made at a time, and not by the instalments. Sometimes government grants were obtained by the affected families, after waiting for three to four years. Loan were distributed to them through the concern "Panchayat Samitee". This work can be done by opening a cell at taluka level by the project.

Due to displacement of affected small landholders many of them have now converted into landless labourers. They are depressed under indebtedness because they lost five years of their agricultural income due to unavailability of land and other facilities related to agriculture. They are now requesting to the government for relief from the loan granted to them of Rs. 5000 to each family. This loan was taken from Z.P. of Satara for construction work of residential house, to each family of New Vyahali and New Chandavadi.

Fixation of Price of the Land

For any irrigation project, it becomes very difficult to the L.A. and Rehabilitation Offices to fix the quality and price of agricultural land from catchment and from command area. It is observed that a landholder of old village was holding 8 acres of unirrigated land now asking to the government, the same acreage of irrigated land from command area. This should be checked by strict and quick management system. Property of land is divided by some rich landholders in the record of village abstract to take the benefit of government facilities for obtaining more land and residential plots. Such malpractices should be controlled by strict punishments.

Price of acquired land is fixed as per rate decided in the year 1930. This rate should be revised with considering the area of lift and well irrigation. The price of agricultural land should be fixed by considering previous five years of standard value. The government rate of land is lower and cheap compared to the present rate of land in the market. For Shri Sailam Project the affected families have estimated value of stone house at Rs. 10,561, for a hut Rs. 2,500, but the government has paid only 5,561 for stone house and Rs. 645 for a hut. Here for Dhom Project the prices of stone house and hut are not properly estimated. Even there are some cases where it is observed that the villagers have paid bribes to related person, for saving the visits to office and not to delay the work in the office.

Sympathetic and Helpful Approach with the Public

The staff members of "irrigation – rehabilitation and L.A. offices have given only assurances of fulfilling all their requirements. When the construction work of the dam is over, nobody turns to help them and they are guided to go to the court for obtaining the demands. It is a duty of related staff-members to guide, help and give full co-operation to the people and give clear cut idea, as per rules and regulations about rehabilitation and the facts should be informed to all the affected families.

Efficient Office Staff

Efficiency of any office is measured by quick decisions and quick actions in related matters. For settling the matters of rehabilitation, the officers should go to the farmers of the villages and should take decisions over the matter. The work of rehabilitation should be done as an emergency work through introducing accurate plans so that displaced-unsheltered families will be converted quickly into stable families.

Rehabilitation Co-ordination Committee

For co-ordination among the government offices, the dam affected persons, and the political leaders the following administrative pattern should be formed to work under the irrigation project:

(*a*) Rehabilitation Co-ordination committee :
Staff for the committee: (*i*) District Officer, (*ii*) Project Officer, (*iii*) District Rehabilitation Officer, (*iv*) Concerned Land Acquisition Officer for catchment and command area, and officer providing all the facilities to rehabilitated village at new location.

(*b*) Rehabilitation Advisory Committee :
Representative members for the committee :
(*i*) Member of the legislative parliament from project affected and benefited area, (*ii*) Two members from Z.P. and two from Panchayat Samittee of concerned part, (*iii*) Ten members from the grampanchayat of affected and benefited villages.

Control on Sale Purchase Deeds

It is found that the big landholders have already collected in advance the information regarding canal bed or canal route passing through the agricultural farms and they started to invest amount of purchasing such land in cheap rates which will come under canal benefited area. This practice should be checked by law introduced against land transition and "sale-purchase deeds" for command area.

Black Rates of the Land

Within the rural part 'land' is an important economic source. If a dam affected landholder wish to purchase land from the command area, he has to purchase the land by black rates which are ten to fifteen times greater than that rate indicated in the record office of Sub-Registrar or in the "sale purchase deed". To save the amount of stamp duty prices of land are indicated in the "sale-purchase deed" so less than the actual amount paid to the owner. The capitalists and rich persons are responsible indirectly to this bad tendency formed in the society as the rich persons always tend to increase the prices of agricultural land. The social movement against this tendency, should be developed within the minds of people who are related to such sale-purchase business. It requires quality of life of the persons in any society. The same experience is observed for "Shri Sailam Project" that the basis of fixation of value of agricultural land, to save the "registration fee", lower rates are generally quoted in the sales deeds in Sub-Registrar's office.

Difficulties and Problems of Affected Persons

The persons displaced are reluctant to leave their old homes and accept to new environmental settlings even though they are offered by the government several facilities and incentives.

Waiting Period for Obtaining Possession of Land

Waiting period for obtaining possession of land resulted over the unstability of their family life. They were disturbed mentally and found unable to enjoy social and cultural functions, festivals, and village annual fairs. From shifting to obtaining of agricultural land and residential plot, most of them have lost generally five years of "exile" period. This period is great loss of their life which cannot be recovered by any way. Some of them have accepted agricultural labour work in the nearby villages of command area, and have taken help from their relatives and some of them were working on a daily wages at the construction work of Dhom Dam.

Once disturbed the village environment by shifting it at another area, eco-system of the village is also disturbed for the next 10 to 15 years. There is only one solution, over this to save from this "loss of period" of their life, that on any way, they should be immediately provided their basic needs and resources in the new villages and equal compensation at one suitable place.

PROVISION OF NECESSARY FACILITIES

To Government Staff

Government has provided all the facilities to the administrative personnels connected to the work of rehabilitation. The staff of the project has provided good residential shelter, electricity, water supply and to their children a school bus facility. Government has constructed good quarters for the project workers, very close to Dhom lake. The settlement was known as "project Vyahali Colony". When construction work of Dhom Dam was under progress the settlement was busy with full liveliness. After completion of the work of the dam this settlement has turned into "ghost region" of lonely houses without people, because all the people have shifted at Satara Kanher Irrigation Project.

To Dam Affected Persons

To some limit it is possible to the government to utilise such constructed quarters or to construct medium small size of quarters for the dam affected persons. It is also possible to provide S.T. Bus facility to the school going children of dam affected villages from their new village up to the school or college area in cheap rates or in concessions.

For generations together, they were emotionally connected to their old environment. They are now shifted and resettled in such a new environment which is with lack of sufficient water and lack of protected shelter for the longer period of two to four years over the infertile land. They have again to spend a period of 3 to 5 years of their life time for rebuilding reconstructing the new environment to bring out a stability and mental peace to all of them. For this purpose at every stage of rehabilitation work, "a plan" should be made and should be implemented by the government for development of the region.

Big and Small Landholders

The big landholders at the new locations have mentioned that they have no complaint against Talathi or Gramsevak of their villages, because they have used their economic power for taking the benefit of obtaining sufficient and good quality of land with the help of the Talathi and Gramsevak. Rich group of households could manage to possess first preference of selecting good quality of agricultural land and residential plot at the centre of the village on the main road.

It is essential to give first preference to the weaker section of dam affected families (Small landholders) at the time of providing land.

THE SHIFT TO NEW VILLAGE SITES TENDS TO INTRODUCE THE PERSISTING PATTERN OF THE VILLAGE RESIDENTIAL STRUCTURE WITH ITS TRADITIONAL DIVISIONS OF CASTE BASED WADAS OR GALLIES

Krishival Village Structure

It was a plan to form and provide an ideal example of new settlement pattern, in which the people of different castes are living together closely within a lane or a "wada". The idea was failed due to "Krishival village structure" and due to traditional religious impact over the rural society. "Krishival structure" of any village should be changed into commercial and scientific agricultural village structure. With the help of experts and educationists it is easy to bring the villagers quickly at rational level.

Absorption of Dam Affected Families

The affected families have obtained firstly land for residential plots and then the agricultural land as per availability of land by the government, resulted that in all the new settlements, residential plots are separated from agricultural farms and there is a vast distance in between settlement and agricultural farms. Agricultural land should be given at the very close to residential plot and the work of giving land for agriculture and for house plot should be done at a time by the government. The aim of rehabilitation should not be henceforth done by forming a new village settlement but by absorbing displaced villagers within the command area villages. A shifted village cannot be resettled at one location. It also depends on the availability of land and willingness of the villagers of benefited area for absorbing dam affected families within the environment of command area. It is also possible to give residential plot within the agricultural farm. This will be resulted to form farmstead type of scattered settlements. But from the study of Dhom rehabilitation it is remarkable to observe that due to social problems of different villages, there is no scope for formation of farmstead settlements. Agricultural farmers of benefited area were not able to mix up and co-operate easily with the new-comers—rehabilitated farmers settled around them.

The mind of the people of command area should be prepared to accept and merge the affected families with them and with their social life because they are benefited by the canal irrigation facility. If they observe against this movement, no facilities of any government schemes should be given the people of command area.

Absorbed rehabilitated village settlements of 2 to 10 families in a group should be planned and settled by providing water, electricity and other facilities and they should be located near the market or industrial centres. Such centres can provide them a good support of different sources of income.

The importance should be given in the rural school for education of local resources. The duty of individuals is to work for the benefit of of society. We should not always consider that any social work is a duty of the government.

Shifting Motivation

Shifting motivation of the affected families was purely economic. It was striking example that children were asking to their parents about why they were shifted so quickly and came to an infertile rockey land to settle. Why our villagers in shoals are shifted away? Why we became homeless and unsheltered? Where are those mango gardens, their rice farms, the windows of their homes, the river, the temples and green beauty of the village? Under sad mood, people could not answer for such questions.

Traditional Village Structure

All the dam affected village settlements were based on traditional oriented, clustering and compact settlements. Due to shifting of dam affected villages they lost their social entity and their community life and houseship also.

Old system of Wada structure is not observed at resettled new villages. Due to lack of sufficient houses newly constructed temple can be utilised for multipurpose work at new location, i.e., for Post Office, medical centres or for school etc. Average land distributed to residential plots to each family is 0.03 ha. (3 Gunthas).

It is a fact that caste system is a unique significance of rural area, largely determines the type of houses, housing habits of family members, and choice of village area. Caste consciousness is stronger in rural population. Village is divided into areas each inhabitated by the member of any caste. It is not possible to form a new settlement in which different community people of different races and religions can be settled in a lane or Wada structure of settlement. Even it is observed in Maharashtra that in all the villages B.C. families are settled away in outside the main village settlement.

In the sccial structure of the villages there is no proper place for the people of B.C. community to settle within the people of developed society. A very critical position is observed when racial and religious conflicts can take place in a settlement formed by the communities of different type of race and religions.

No B.C. families are observed ready to settle near the temple of the village or close to the settlement of Maratha and other developed families. At new Vyahali and New Chandavadi hence B.C. settlements are located away to the south-east from the main village settlement. Still caste based human relations resulted to divide people or rural area in different communities.

Instead of settling a separate new village, there should be twine settlements formed near the benefited villages. "Rehabilitation" should aim at merging affected families in different villages of command area and separate new villages should not be formed. Five to twenty families can be merged within different villages by giving them residential plots and agricultural lands. It is a social binding for the villagers of command area to accept the new-comers within their villages. They should not take the new families to be enemies only because they have acquired their land from benefited area.

The landuse pattern in the new villages tends to be different from the one in the original villages under the influence of the new environmental and economic setting.

At the old locations, most of the farmers were accustomed to dry farming. Rice was the prime crop within the valley. Agriculture was fully depended on monsoon rain. Rice, jowar, bajra, nachani, was their agricultural productions. Western part of Wai Taluka, along the course of the river Krishna including the catchment area of Dhom Dam was once covered with dense forest and grasslands. The entire area has been slowly deforested since then.

Occupations at Catchment Area

The occupations of the affected villages were based on primary activities supported on local resources, i.e., agriculture, forests and grasslands. Working as a Mali in the private gardens, making sticks, collecting honey, preparing rcof of the grass to the buildings as protection from rain in rainy season were the important occupations to dam affected villages of the area provided by the closed environment of Panchagani hill station. The catchment area of Dhom Dam was also well known for rice production. Rice was a popular crop of the area. Dairy farming was important source of income. Now at the new location of the villages, due to unavailability of sufficient grassland, most of the landholders have lost their dairy animals by selling them elsewhere and strength of dairy animals is reduced by 50 per cent of the villages of Chandavadi and Vyahali. At the old villages cattle were easily reared within the grasslands of nearby mountain slopes from the private and government lands. Poultry birds were reared on gutters and ditches of manure. The villagers did not care sincerely for the health and life of their domestic animals and birds.

Scope for Tourist Industry

At the vicinity of village Pasarni, on Wai-Pachagani road there is a well known silk centre which can support central part of Wai Taluka for sericulture. This part is also suitable for flower gardening and honey making industry, and for cultivation of medicinal plants. Central and western part of Wai Taluka of the upper Krishna Valley is prosperous area in the future for tourist industry, due to availability of beautiful natural sites (i.e., Menavali, Dhom, Wai) and temples along the river Krishna. Dhom lake is the new attraction for the tourists while passing through the road from Wai to Pachagani (from Wai on 13 km. distance) and Mahabaleshwar (on 32 km. distance)—hill stations of Maharashtra.

Benefit of Developed Part in Command Area

All the villages rehabilitated by Dhom Dam were located in the very backward part of the region. The villages were connected with their relatives in the nearby old villages. There was hardly relations of the dam affected villagers with the central and eastern part of more developed Wai Taluka, so all the Dhom Dam affected families have forever lost their villages, agricultural land and also their relatives of nearby villages at old locations under catchment area of the dam. They are now shifted and located to more developed part of the region at different places in different Talukas of Satara District. They should take benefit from this developed part for increasing their quality of life and they should bring economic stability to their life through other source and the source of irrigated agriculture. Now, shifting resulted in their occupational structure from rainfed farming towards irrigated farming region. In the new environment sugarcane, wheat, vegetables are cultivated. Change of cropping pattern from rainfed into commercial farming is resulted over the yield of crops. Modern agriculture needs a strong industrial base. Rural skill should have to be generated in the local fields by planning man-power in agricultural occupations.

Scope for Agro-based Industries

Here is a list of different type of occupations based on local resources and can be introduced in the rehabilitated villages of Krishna Dhom Dam.

(*i*) Sericulture (Silk Industry), (*ii*) Fruit and flower gardening, (*iii*) Honey and stick making, (*iv*) Oil industry from sunflower, groundnuts and other oil seeds, (*v*) Poultry farming, (*vi*) Dairy farming, (*vii*) Grain threshing, (*viii*) Preparing Mats, (*ix*) Hand made paper industry, (*x*) Making ropes and purses from sisal, (*xi*) Preparing paper bags, (*xii*) Co-operative unit shops providing seeds, fertilizers and other materials in cheap rates, (*xiii*) Preparing soil pots, (*xiv*) Sewing and knitting, and (*xv*) Jaggery making industry.

Present Occupations of the Project Area

At present there are following occupations and industries located nearby in Dhom Project area :

(*i*) Satara Sahakari Sakhar Karkhana Ltd., Bhuinj (Sugar Factory), (*ii*) Silk Centre, Wai, (*iii*) Tata Perfumed Oil Plant, Pasarni, (*iv*) Hand made paper industry, Wai, (*v*) Oil Mill, Pachwad (not in working condition), (*vi*) Clay model industry, Wai.

Multi-purpose co-operative units should be organised for uplifting the needs of villagers for better farming and better living.

There is no sufficient speed for establishment of cottage and small scale agro-based industries in the studied rural area due to lack of knowledge of utilisation and conservation of local resources and lack of manpower planning. Most of the rural energy is consumed under traditional and backward occupations and for political movements, which are managed by unqualified and selfish persons and resulted to unqualitative products. Sometimes prepared better plans and research methods remains in the books. Implementation of such plans is necessary.

No Possibility to Prepare Landuse Map of the Affected Families

It is not possible to prepare a villagewise landuse map of studied area, due to agricultural land is obtained by the affected families at different villages at different locations, like patches.

The Lower Part of the Catchment Area and the Big Landholders

The fertile land of lower valley at the old village area lost under submergence of Dhom Dam was occupied by most of the rich landholders. The infertile upper part of the valley was covered with small patches of farms by small landholders. There are some families (30%) observed that for the want of money and to solve the unexpected difficulties (like to pay medical charges of sick person in the family or to pay marriage expenditure of the member of the family) they have mortgaged their agricultural land to the big landholders.

Pressure of Population Over the Land

Villagers of command area along the Krishna River Valley are settled very close to each other. As per growth rate of population, the pressure on land has increased. Farms have shrunk to one acre and even less than 20 Gunthas. Land fragmentation becomes more accute resulted over for formation of small landholders, which are not good farmers due to poverty. Land should be consolidated in large size of farms, for self-sufficiency in food, we should have to attain per head 5 Gunthas of irrigated land.

Necessity of Water Management

It shows that due to canal irrigation facility, the command area people will incline to cultivate cash crops to obtain more profit. From the command area of Krishna Dhom Project the land under sugarcane cultivation is increasing rapidly. It should be remembered that the water supplied to 0.04 ha. (one acre) of sugarcane can be used for the growth of 4.04 ha. (10 acres) of oridinary food crops. The sugarcane consumes about 8 to 10 per cent of the water supply.[2] Cropping pattern in the new canal irrigated land should be scientifically arranged by considering the quality of soil, quantity of water available with climatic conditions and local requirements of food and cash crops, water should be utilised very carefully in the command area of any irrigation project.

No Choice of Selection of Agricultural Land to the affected Families

Poor and small landholders of affected part have explained that they had no choice for selection of land in rehabilitation. The only way before them was to accept and agree to settle as per availability of land at any village area fixed by the Rehabilitation and Land Acquisition Departments. The village artisans have left their traditional occupations and have accepted agricultural work as labourers, except Mang, Chamar, Nhavi, and Gurav, doing traditional occupations and agricultural labour work on daily wages. Without consideration of quality of soils, the affected persons have accepted land at rehabilitation area from the government. This is because they once refused to take possession of the land, they were afraid that they would remain unstable for a long time in future due to unavailability of land with the government. Those who held infertile land at old sites, have tried to obtain fertile agricultural land at new sites. This practice should be controlled by keeping up-to-date record from catchment and command area. Quality of acquiring land should be recorded carefully and a policy should be adopted by giving land in the place of same quality of land as compensation. Land for land is the best policy in this connection. If infertile land is given to affected persons, proper methods should be introduced for increasing fertility of soils. The land can be used for cultivation of Sisal, which can check the soil erosion along the slopes of the region and can yield raw material for occupation of preparing bags, purses and ropes from sisal.

Majority of families have obtained agricultural land at a distance of 10 kms. from the new village settlement. A peasant has obtained in small patches of farms at different places and he has to walk this distance twice daily. The affected persons have never asked the government to give residential plot within the agricultural land. In this case also it is observed that they were afraid of 'rejecting the land' would result unstability for their life in future. Government has acquired at first the area of settlement and then the agricultural land from command area.

There are some affected families of New Vyahali, which have obtained agricultural land at village Kalambe, Tal. Satara, and they are residing at new Vyahali, Tal. Wai. How can then we expect growth of agricultural production from resettled villagers if their land is far away from the new settlement?

THE OVERALL EFFECT OF THE SHIFT SEEMS TO CHANGE THE PATTERN AND LIFE STYLE OF THE PEOPLE WHICH IS DIFFERENT FROM THAT IN THE OLD VILLAGES

Social Fabric of the Affected Villages

For the villagers of Kadegaon and Khadki (from command area) the new arrivals of Vyahali and Chandavadi feels as parasitic and unnecessary burdens. They could not maintain emotional entanglement. A social fabric of the affected villages is destroyed due to rehabilitation. Government could not form unity and emotional entanglement within the displaced and canal benefited villagers. Social security is diminished of the dam affected families as they are scattered shifted at different places. The coming generation from them will lose social integrity and social consciousness. Rural life of this area is destroyed due to push and pull factors. Dam affected villagers are shifted and pushed compulsorily and resettled pulled at suitable locations in command area.

Due to shifting of dam affected villages, we have divided and scattered a compact society at different new locations. We have disturbed geographical base of the region. We should have maintain geographical equilibrium by resettling shifted villages and their social and economic life at new locations. Resettlement should not be worked out by only economic motivations, but also with the cultural, social and with natural developments. Due to present system of rehabilitation, it is found that a social fabric of affected villages is disturbed and ultimately destroyed. A consciousness should be developed within the rural people to work and plan for the progress of community.

Religious Traditions

Annual fairs, Bhajans, Pujas, and other social functions and festivals are continued in the new villages and being traditional many of them have shown interest to participate in such social functions. The base behind of such functions is religious. The statues of village deity or god are also shifted at the new locations. The investments in annual fair, Satyanarayan Puja, and marriage ceremonies are celebrated by many villagers with the help of taking loans from the possible facilities available in the area or from the "Jatra-Fund". Before the day of annual fair a part of "Jatra-Fund" social amount is divided into Rs. 50, 100, 200 or 500 and distributed to needy persons or to the poor villagers by auction which is recollected in the next year. Religious traditions are so hard and rigid that nobody wants to change them. Sciences are elastic but they are not entered in the minds of rural people to change the rigidity of rural occupations and traditions which are harmful to the society.

It is also found that the villagers have habit of killing their time for discussion of political leadership of the village and every leader of each group, shows himself as a great political thinker, resulted on forming two against groups in the village politics. It spoils the village development plans, due to their shallow and selfish nature of behaviours. They should understand to use their mental power for the constructive work of their society. Lazy and inactive persons, should be awakened and generated for conservating and cultivating local resources.

Migration at Long Distance

In the past migration on long distances by the family members of rural area was restricted at the nearby villages, but when the problems of poverty becomes more acute, the villagers have started to migrate anywhere for getting employment jobs even at a long distance at Bombay, Pune, etc., within the state.

Diversified Life and Empty Errants

To the dam affected families we have given lands but not the resources. Family life of all the people has been diversified. Many of them have obtained insufficient land in the process of rehabilitation. They have wasted their time, energy and amount for obtaining government grants sanctioned to them as a compensation of their land by visiting more than six times from one to another office. They are tired with these empty errants to different offices. Due to their illiteracy, they do not know the proper offices in relation to their work. No office staff has helped and guided them properly to solve their difficulties. It is observed by them that everybody from government offices has a habit to return and put the ball to the other side and of pushing and shouldering the responsibility with other office. Society should move and come forward to take some steps against such matters.

Help from Benefited Society

It is also the responsibility of rural society to help and to form a new life in the new environment of the rehabilitated sites. It is not individual work. The affected persons were expecting a great support and help from those persons who are benefited by canal irrigation. A small part of amount collected for annual social fairs from command area villages can be given as a token of love and such little gift may be a great givings if it is given with love to the affected persons. Rehabilitated persons, get adjusted-mixed with new environment very easily. They could do so, due to no other alternative. The affected persons should be provided different agro-based occupations from the investment made and incurred by the government or they should be utilised for development of local resources of the region. The ultimate aim of this is that the displaced families can maintain and fulfil their needs. If a land is acquired for canal, alternative land should be provided.

Marriage System of Dhom Catchment Area at the Village Borgaon at Old Site

Early marriages are common in the village Borgaon. Before marriage the girls can reach upto the primary examination of VII standard. The result of this is that girls are rare for taking benefit of Secondary Education. Borgaon High School has opened VIII class in 1982 and IX Class in 1983. There are only five girls taking benefit of secondary education. From the year (1984-85) the Headmaster of Borgaon High School proposes to start residential facility in the school.

In this area there are 90 per cent of people living from Maratha community. In this community there is another tradition of practice of remarriage under the pretext of no issue from the present marriage or the female giving birth to female issues (girls) only. Some instances of remarriages to the extent of 3 to 5 in number were observed, in the village Borgaon. Instances of remarriages of the divorced females having a good record of child bearing were also noticed.

There are educated youngsters in the rural area who strictly follow of the traditional path of taking "Dowery in marriages", otherwise they are not ready to accept marriage settlements. All the families of rural area of dam affected and benefited part are still following the same traditional path of old marriage systems.

Education Facility at Catchment Area

The submerged area of Dhom Dam in the upper Krishna Valley comes under mountainous and economically poor part of the region which resulted backward population in education and very few graduates are found in the affected villages. At present there is only one High School working at the old location of partly affected village Borgaon in the valley. Students come to the school barefooted. They cannot afford to buy enough clothes, books and other things due to poverty. They take insufficient food and walk 6 to 10 km. distance to attend the school. Since past generations this is the first time that the poor boys could step into the life of secondary school in this area. "New English School, Borgaon" (High School) started its working from June 1980. The Head Master is the founder member of this School. Borgaon, Dasawadi, Chikhali, Malatpur, Dhavali, Dahyat, Waigaon, Jor, Golewadi, Nandagane, Partwadi, Akoshi, Korande, Nhalewadi, Mugaon, Kondavali Bk. and Kh. are all the feeding centres to this high school. The school is serving to all partly dam affected villages. There are some students coming to school daily from a distance of 11 kms. from Nandagane and Partwadi. There are 7 male teachers working in the school and for the present (1982-83) the strength of the Secondary School is 210 (175 boys and 35 girls). Post Office, Electricity, Family Health Centre, S.T. Bus (From Wai) are the facilities available in the village from the year 1982.

Previously in this region there was only one High School at Asgaon which is completely shifted near Vaduth, Tal. Satara. The challenge to work for education backward people of this area is accepted by the Head Master of Borgaon High School.

As per Borgaon High School working for Krishna Dhom Dam affected students, there is also in the Koyana Valley, at Taldeo (Mahabaleshwar Tahasil) a residential High School founded by Shri. Tanaji Babar, the Head Master of the School who has taken a lot of pains for developing the school for the backward people of Taldeo area which is the affected region by Koyana Project. Mr. P.L. Deshpande, a well-known Marathi writer has donated Rs. 25,000 for the development of school laboratory named as "Kanad Prayogashala". In such region it is very essential to start residential school for the affected villages.

Educational Facilities to the Affected Villages at Command Area

Now at the new locations of studied villages of Vyahali and Chandavadi, Higher Secondary Educational facilities are available at Bhuinj for Chandavadi and at Pachwad for Vyahali. For them Wai is the only centre for college education. "Kisan Veer Mahavidyalaya, Wai" provides college educational facility to Khanadala, Jaoli, Mahabaleshwar and Wai Tehasils of Satara District. It is the only one profit that Dhom Dam affected villages are rehabilitated in the developed part of the valley from their

"Tyajet Gramam Janpadasyarthe"

1. FOR THREE GENERATIONS..................
.........Future.........Stable.........? ....
or unstable.........? ?? "Rehabilitation due to irrigation project : is a man made disaster"

2. "We can't recall the past, but, we can reconstruct the present."

3. The prosperous tourist region : around the Dhom Project :- **WAI** Taluka Place : (Distance from Dhom : 9 km to SE)

4. **Menavali :-** (Distance from the "Wada of Nana Phadnis Peshva." The walls are carved and coloured with pictures showing "The great Peshvai Art."

5. "Menavali Krishna Ghat."

6. **Bhogaon :** (Distance from Dhom Project: 3 Km. East) (Samadhi Place of well-known Marathi poet of 17th Century : Shri Waman Pandit). At present, the village comes under command area of Dhom Project : The left bank canal near Bhogaon is shown in the photograph.

7. **Village Lohare :-** (Distance from Dhom Project : 15 Km. to NE).

8. Southern part of the dam wall, where old Vyahali was settled.

9. Pandavgad (fort) : from the top of the Dhom - Dam - Wall.

10. The Dhom Project (from South East): A new life for the villages of command area.

12. The Dhomeshwar Temple of the Village Dhom: (Old location) The temple is still remained unaffected from Dhom Dam.

11. Dam Wall of Dhom Project : Southern part.

13. Village Dhom : (Old location) Now comes under "dam protected area." (a) A narrow lane-compact settlement of Dhom.

14. Road made by cobbled stones, connected the village with Dhomeshwar Temple.

15. Gomukha (near "Dhomeshwar Temple) : a source for supply of drinking water to the village.

16. **Village Dhom : (Old location)** A thatched-roofed house connected with closed cattle shed.

17. The project Director / the author, delivering a lecture on the occasion of "Annual Social Gathering of "Narsing High School, Dhom" on 8th March 1982. ....and now we have to reconstruct such—environment at the new locations.

18. The new location for village Dhom at Joshi Vihir with in Ozarde village area.

19. Roads are marked with lines. This photograph is taken from Pune-Bengalore Highway which passes from the East of the new settlement.

20. A house of old Vyahali now lost under the water reservoir of Dhom Dam. This photograph was taken when the settlement was waiting for merging under the encroachment of stored water of Dhom Dam.

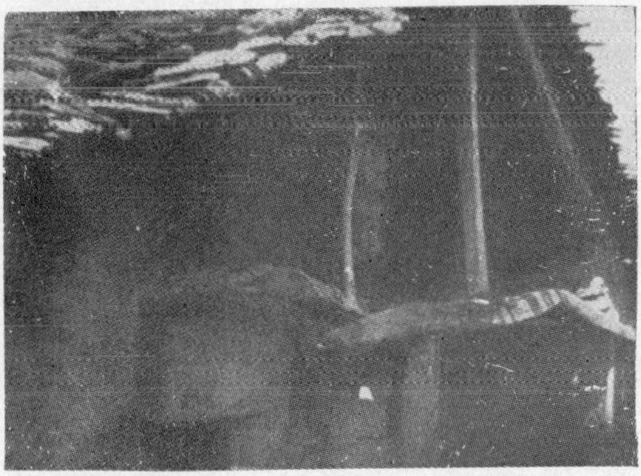

21. Site of old Vyahali :- The mountain range of Pasarani Pachgani Ghat sloping down towards north at Dhom Lake.

22. Three houses are observed in the photograph, still remained unaffected due to their part of old location (old Vyahali).

23. The catchment area of Krishna Dhom Project was well-known for Rice cultivation. Rice threshing through bullocks was common practice within the area.

24. A part of the settlement of old Vyahali.

25. **The natural beauty which for ever lost under the "Dhom Lake."**

26. Western part of old Chandavadi.

27. **Old location of Chandavadi :** This photograph is taken in summer, when the level of water of the dam was very low. Top of the tree, indicates old location of village Chandavadi, drowned under Dhom Lake.

28. Dhom Lake, Walki & Krishna "Sangam", separated by a spur.

29. Old location of a well-known cattle market outside village Velang washed out and lost under the waves of the water of Dhom Dam. Now the area is covered by sand and mud. Water of the dam is seen in the photograph.

30. **Location of New Vyahali at Village Kadegaon area.**

31. New Vyahali from Satara-Wai road.

32. New Vyahali : Sheltered by "Virat Gad" to the south of the settlement.

33. Central part of New Vyahali :—
The "Bhairav Temple and the play ground of the primary school in front of the Temple.

34. The eastern part of the new settlement. In the background is the slope of "Asale Bhavani Temple" hill.

35. New Vyahali :—A small well providing drinking water to the village. The well is located at the backside of "Bhairav Nath Temple."

36. Central part of New Vyahali

37. Unsettled part of New Vyahali and Wai-Satara road.

38. The Eastern part of New Vyahali: with an agricultural farm.

39. A lane in the western section of New Vyahali.

40. Right Bank Canal to the south of New Vyahali.

41. New Vyahali : B C. settlement to the southeast of the main village.

42. Source of canal water : closed agril, land of new settlement, Vyahali

43. NEW CHANDAVADI (at Jamb Mal near Bhuinj). From Bhuinj, on the way of Chandavadi "The Satara Sahakari Sukhar Karkhana Ltd." (Bhuinj Sugar Factory) Kisan Veer Nagar is located.

44. This is the barren land which is selected for rehabilitation of villages Chandavadi, Velang, Renavale. Ancient Mahadeo Temple is shown in the photograph.

45. On the way from Bhuinj Sugar Factory to new Chandavadi, the following villages are rehabilitated at one location : Bhivadi Vasole, Renavale, Khavali.

46. "Welcome" to New Chandavadi : This is the eastern rocky & barren land of village Khadki (Tal : Wai),—locally known as Jamb-Mal" - at where—Chandavadi is resettled.

47. A lane of New Chandavadi : The building with a pole of electric wire, shown in the photograph is a building, with small stationery and kirana shop of the village.

48. A pacca house of New Chandavadi, with cattle shed.

49. New Chandavadi : Shelter is constructed by a covered roof of tin sheets.

50. New plantation of trees along the road of Chandavadi.

51. Newly constructed shelter : Outside part is shown with cattle shed.

52. Temporary shed is constructed for the temple of New Chandavadi. The constructed water tank is not in use at present. (May 1982)

53. In 10th May 1982, there was a marriage ceremony at New Chandavadi. The photograph represent - the "Bridegroom" entering in the village by a procession of marriage party called in Marathi as, "*Lagan Vanhad*"

54. A lane of New Chandavadi.

55. Eastward part of the village settlement : The agricultural farm is shown with sun-flowercultivation.

56. Primary school of New Chandavadi.

57. New Chandavadi : Water supply through water tankers to the village.

58. New Chandavadi : Water Supply through water tankers to theVillage.

59. Menavali : Benefited village of Krishna Dhom Project - Left Bank Canal.

60. In the Upper Krishna Valley, it is common that all the B.C. settlements are located to the Southeastward and at the outside part of the main village settlement. B.C. settlement of village Lohare : This is a rare example of showing very closed settlement of B.C. community, divided by only a road with the main village settlement.

61. A market centre of the area.

62. Vyahali colony : Well equipped Government quarters for the staff members, working under the different offices related to "Dhom Dam."

63. A farm of village Menavali : Dhangars from Aire Tal. Khandala coming with their—shoals of "sheeps and goats"—for every season (from October to May) at Wai region.

original backward area. Bare mountain slopes of the new village area should be used for cultivating sisal and other forest products.

To all the students of affected families should be provided concession in educational fees. S.T. bus passes and educational materials, books in cheap rates by the educational institutions.

In all the affected villages, female education is completely neglected by the rural society. The strength of uneducated female is more due to having a long distance of education facilities, from their settlements. Hardly a few of them can reach up to the college education. How can then we expect scientific and rational approach towards life of such daughters who have not been benefited by Higher Secondary and College education? To educate a girl is to educate a family.[3] Unfortunately education for girls has not made progress in our country.

Education and Rural Society

Rural society does not seem to have any sort of interest of reading good quality of literature. Library facilities are not reached up to the villagers through proper ways for maintaining interest of reading economic, social and agricultural books. The agricultural sector of rural part lags behind due to non-entrance of educated skill and adversely resulted in lower productivity. This referred to small and medium landholders. Having a little education, the youth prefer to be a peon or a clerk, rather than to work in agricultural farms. In cities, a student of science can prepare a model of any scientific instrument or any scientific experiments as per his knowledge. Why then in rural area, a student cannot attract towards his agricultural farms for making experiments with soils to get profit from the land? To cultivate interest in agriculture and in agro-based occupations within the minds of young generation, our primary and secondary education should be with a special subject of "Agriculture: practical and theory", for rural school. We have sufficient Primary and Secondary Schools but we never stressed education for Agriculture: (water management cropping pattern, soil survey, development of natural resources, etc.) in rural schools. Small and medium landholders are unable to fulfil the requirements of land, soil, crops of their small farms. They are rich in physical energy but very poor in economic power. Poverty keeps them unproductive in agricultural farms. Newspapers are good agents of distributing knowledge for the people. But in majority the Marathi newspapers are used mainly for political news. They are failed in stimulating scientific agricultural knowledge and are observed unsuccessful to represent the social economic problems of rural regions in the studied area.

People of rural area should understand to keep equilibrium of nature and resources with population. Nature is unselfish. Nightingale sings without an audience, sun gives insolation without considering the race and religion of the people. Trees produces fruits and wood for others. Man takes many things from nature but he gives very little to the nature. Nature and resources should be utilised and developed very carefully for satisfying unlimited wants of human beings. In this method people should be stimulated for utilization and conservation of natural resources. The society is formed with different type of economic and cultural colours. The colours should be attractive, closed to each other and stands in a belt unitedly like a rainbow. There is a particular discipline in the nature, but there is no discipline for man in a society of rural part. Rural society is behind the economic satisfaction. Cultural qualities of human life for mental peace are not properly maintained in rural society. For this work of organising seminars for the society should come forward. It is the duty of society rather than a person, to take interest for participating in it. It is tragedy to note that the studied rural youngsters are in general not sincere and are not able to understand their present and future problems. They are unable to work for development of resources. They do not have the habit to work for others.

Rural illiterate person is not ready to accept the reality around his environment. He is twister and he kills his life without principles, and changes his opinions without checking the reality on any matter. He is boastful, and unable to take proper decisions on any matter. Sometimes he is observed weak in mind, and accept anything might be unhuman and not useful to his society. He does not easily accept the expertised views put forth to him. This social position is resulted over backward type of agriculture and less productivity from it. There is no use of only in the growth of agricultural production, if we could not develop growth of human qualities in rural society. Dirty and unhealthy traditional environment resulted for growth of diseases and even they have not time to think and understand the value of "cleanliness is godliness". Mean minded people always trying to bring obstacles on the path of cultured and developed families in rural parts. The villagers are satisfied with the position prevailed by the grace of god. The tendency of "ठेविले अनंते, तैसे चिरहावे, चित्तीअसो द्यावे समाधान" is hampered to cultivate rational and scientific outlook and attitude in their minds.

There is a vast gap in between the old and new generations in the rural area and any social development plan suggested by the youngsters is never accepted by the old generation.

Youths from rich families of landlords are found lazy and observed unwilling to develop skill in academic and technical fields. Even they do not bother for the growth of culture and quality of human life. Females in the rural area are not able to participate in social and co-operative fields of their villages,

They are meant for household duties. The women are not supposed to cross thresholds of their houses and thus they remain dominance of men. It is found difficult to marry off the daughters of dam affected persons due to unstable life in the new environment. Energy of females should be planned for utilization of different occupations (*i.e.*, sewing, knitting, etc.).

Illiteracy resulted over the growth of dependents in the rural families. A family with ten members, five are earning for ten, the budget is maintainable. But if out of ten only two are earning for ten, the budget is not maintainable and adjustable to the family. In this position, members of the family are forced to leave the village to seek employments. It is necessary to co-ordinate population with resources and regionwise resource development schemes should be organised by man, for welfare of man and nature.

From the studied villages, it is found that one-third of earning population is only responsible for livelihood of the remaining dependents (two-third population).

Habit of Keeping Accounts and Showing Annual Income

It is observed that many villagers do not have habit of keeping accounts and family budgets because many of them belongs to the category of small or medium landholders and they do not get profit from their limited agricultural lands. It is a tendency of the studied village farmers that to show very little annual income than the actual income of their families for taking benefits of government schemes which are useful to low income farmers. At the time of settling marriages of their sons, they try to represent themselves as rich as possible to the bridal party. It is just like a tendency of women to show their appearance as young as possible and indicate their age 3 to 5 years less than that of their actual age. Festivals are celebrated by some villagers by taking loans for showing their riches. They keep their poverty in the dark.

The people should be raised on such rational level to accept the reality around their environment. This will be successful by introducing workshops and seminars of "non formal education" to the rural society. They should be exposed and utilised their energy and strength for constructive work.

DAM CONSTRUCTION INTRODUCES CHANGES IN THE ECOLOGICAL SETTING BOTH IN THE CATCHMENT AND COMMAND AREAS, THESE CHANGES NEED TO BE TAKEN INTO ACCOUNT FOR THE PURPOSE OF PLANNING AND DEVELOPMENT IN BOTH THE AREAS

Old Environment

The dense forest cover and major fertile rice farms of the catchment area is disappeared and lost under the water of the Dhom lake. Now scattered trees and bushes of karvand, jambhul, mango are remained within the upper part of the Valley. Rice farms in small patches are observed closed to the area of partly affected villages. These settlements, now becomes lonely and depressed—scattered settlements, again separated from each other by the stored water of the dam extended within the Walki and Krishna Valleys. Borgaon is an important village and service centre to the area providing primary and secondary school, Post Office and other facilities.

This region comes under "heavy rainfall", can be utilised for cultivation of jambhul, tuti, karavand, teak, mango and many other such important medicinal plants as well as the barren mountain slopes should be developed under pastures for cattle rearing purpose and by introducing "terraced farming", under rice cultivation. At the closer part of the stored water of the lake, will be in future, important tourist region and for that the area should be developed by utilising local manpower for cultivation of fruits and flowers. Boating will also be an important attraction to the tourist visiting to Dhom dam, which comes at the closer environment of Panchagani hill station.

New Environment

The picture of command area is rapidly changing from dry farming to productive—canal irrigated agriculture and economic growth resulted stability of the people. Vegetables, fruits, turmeric, sugarcane are the new crops cultivated in the region which was originally barren land or under jowar cultivation. It is essential to cultivate economically important trees in the region. Availability of canal water throughout the year, should be managed carefully within the command area—agriculture. The region will be very important if planned development of local resources and agro-based industries should be introduced and managed by the people, with the help of the experts. Command area environment becomes now as "sugarcane environment, due to extensive cultivation of it, along the right and left bank canals of Krishna Dhom Project, which is supported to the "Bhuinj Sugar Factory", of the region.

Implementation of Plans

Rehabilitation of persons displaced by a project is the responsibility of government who should accept physical rehabilitation by providing them similar or near similar land houses elsewhere. A scheme of dam construction and rehabilitation should be worked out simultaneously. The work of rehabilitation must be completed before the lands of catchment area are actually submerged under the water of the dam. Here is an example of Kalinadi Project,[4] from North Karnataka District, to be completed in 1984. In the first stage of the "Kali Nadi" project, the problem of submergence is sought to be solved by creation of a new township of about 20 kms. away from "Ramanagar", where 5000 people from 17 villages have been shifted. For the result of fruitful resettlement in new created villages all the possible occupations should be introduced, for example horticulture, poultry, dairy, sericulture and many such occupations can be helpful to improve the economy of affected population.

Irrigation facilities are increasing considerably but water from dam has not reached to the farms of poor farmers. Improved seeds, fertilizers and timely credit should be made available to small and marginal farmers. The poor farmers should become producers from their small agricultural farms.

In general it is found in the studied part that there is completely absence of consciousness for cultivation and conservation of trees. "Family planning", "grow more food", "grow more trees", "work is worship", "work for betterment of all" are the slogans, accepted but not actually implemented by many of the landholders. It is a high time of not to say but to work for plans in the actual fields for obtaining benefits for all. For future prospects, for benefit of future generations and for welfare of man and nature a social discipline must be compulsorily accepted by the member of the society as a honourable work.

We construct irrigation project and project affected persons are compulsorily shifted away. Why then there is no compulsion to the people and by the people, for cutivating plants in the area, for family planning of not more than two issues, for conservation of forests, for protecting man from air-water-sound pollutions, and for such many plans which are useful for society and ultimately benefited for all the people? To shift a village is a must from affected part of the project. In this way to have birth control by family planning is also a must, not an inhuman deed should be accepted by everybody.

Planning and Imbalance of Unequal Distribution of Public Amounts

Government agencies always try to represent the statistical data of about the growth of any resources but such data are not carefully collected due to lack of mature field-investigators. Accurate measurements of available resources must be collected for planning in the area. Without considering the resources of the area, personal interest of local leaders is badly affected over any plan by pulling the grants towards each other which is resulted imbalance of unequal distribution of public amounts in rural area. Sometimes plans remains on paper and in the lengthy filing process of present administration. There is no necessity of celebrating "Revenue week for gentlemanliness", "Soujanya Saptah" (सौजन्य सप्ताह) by the revenue departments. It is their responsibility and duty of the work for which they are paid by the government. They provide important documents and certificates, but unfortunately it is observed that the office workers are behind money and to take bribes is a common practice. "Surprise visit" by supervisor, must be necessary to stop such bad practices in any part of departments. "Celebration of Soujanya Saptah" is indirectly accepted that the remaining period they are not working by giving kind co-operation for the work with the public. This has been explained by many villagers of the studied area.

Prize System in Administration

The base of any planning should be formed by dividing any area into geographical regions. Suitable geographical locations of the region of growth centres should be developed as administrative centres. "Prizes" and attractive "gifts" must be declared by any office for the best workers.

Rearrangement of Manpower

Eighty per cent of rural population, depending on agriculture, should be reduced up to 25 per cent and the remaining manpower should be properly managed in agro-based occupations or in other department fields. There should be separate co-operative societies for the weaker section of the community and concessions should be given to them by government. The village society should be multi-purpose unit to change Krishival village structure into agro-based occupational scientific agricultural village structure. Growth of population is burden to the country but growth of quality of population is a valuable resources for the country. Culture and quality of human life should be raised for the betterment of our country.

Generally all the affected landholders of "Krishna Dhom Dam" are observed self-centred and "struggle for existence" in the new environment is a challenge before them. The developed and economically stable families should come to help to the uprooted poors by "Krishna Dhom Dam". Urban rich persons should take a benefit of social work by helping to the dam affected persons.

Working Hours and Climatic Conditions

Nature of the work of skilled and unskilled energy of man should be always decided on the basis of climatic conditions in the region, so that fruitful results can be obtained in the quality of work and production. The working period of any industry or office, should be reduced in such a way, so that unemployed youngman power can be utilised and managed to work for the remaining period. This will help to improve for obtaining results of quick administration. Working hours should be decided according to seasonal climatic conditions in India.

REFERENCES

1. Zimmerman Book "*Irrigation*", page 13.
2. Namjoshi, M.V. and Sathe, M.D. (1978), Planning for Weaker Section, Centre of the Studies in Social Science, p. 172.
3. Kabir, Humayun, 1967, Women's Education in the Rural Area: A Symposium, Publication Division, p. 255, New Delhi.
4. The Times of India, Newspaper, March 13, 1982, p. 13.

Appendices

APPENDIX I

Questions in Relation to Rehabilitation and Answer from Dam Affected Villagers

(Vyahali 32% and Chandavadi 39% sample families are studied for the project)	Vyahali		Chandavadi	
	Yes	No	Yes	No
1	2	3	4	5
1. Are you settled in your farm house in the obtained agricultural land?	5	30	7	31
2. Have you obtained agricultural land as per your choice?	8	27	9	29
3. Are benefited villagers in command area helping you for agricultural implementation work?	7	27	11	28
4. Are the village Talathi and Gramsevak giving you a good help and co-operation in time?	14	21	24	14
5. Co-operation from village Patil and Sarpanch	26	9	28	10
6. (*a*) Rehabilitation is benefited	9	26	2	36
(*b*) Do you dislike the rehabilitation	12	23	5	33
7. Is there anybody who have taken cognisance of your problems?	4	31	10	28
8. In the work of rehabilitation have you cheated or deceived by anybody?	7	28	3	35
9. Have you taken interest in tree plantation in the new location?	5	30	19	19
10. Have you obtained agricultural land and residential plots at one location?	6	29	8	30
11. Are the families residing near to you as they were at in old village?	19	16	22	16

APPENDIX I—*Contd.*

1	2	3	4	5
12. Are you landholder of your old village society?	1	34	13	25
13. Due to rehabilitation your family life is improved?	2	33	1	57
14. Have you taken dowery in the marriage of your son?	11	24	13	25
15. Being a dam affected person are your relatives ready to help and to keep for maintaining family relation with you?	11	24	18	20
16. Is it possible to maintain your family budget at the new location?	6	29	12	26
17. Do you work in your own agricultural farms?	28	7	15	23
18. Is there anybody from your family taking interest in social work? (The meaning of social work is taken by them as political work).	9	26	10	29

APPENDIX II

Krishna Irrigation Project : Dhom Dam at a Glance

1. The villages affected by the project : 29.
2. Partly affected villages : 15.
3. Fully affected villages : 14.
4. Total population affected by the project :

	Male	*Female*	*Total*
Fully from affected	3655	5590	10245
Partly from affected	3587	3952	7539
Total of all the affected villages	7242	9542	17784

5. Number of households affected by the dam :
   Fully affected households : 1543
   Partly affected households: 1350.
6. Landuse of original parts under Dhom Dam submergence :
   Crops: Rice, Ragi, Jowar.
7. All the villages were available with Primary School facility.
8. Irrigation source for original location : River, Wells.
9. For drinking : River, Wells.
10. Occupied residential houses : 2541. (were at original location).
11. Total area lost from the villages : 24396 acres
    Fully affected : 11738 acres.
    Partly affected : 12658 acres.
12. Irrigated land lost by the affected villages :
    Fully affected By wells : 393 acres.
    By rivers : 1271 acres.
    Partly affected By wells : 509 acres.
    By rivers : 997 acres.
    Total irrigated land : Wells Rivers.
    902 acres+2268 acres=3170 acres.
13. Out of above 29 villages there were 7 electrified villages.
14. Difference in between old and present landuse :
    Rice cultivation were along both sides of the river as an important sources for rice. The other rain-fed crops were Jowar, Ragi, through oil engine water of the river was lifted up to rice farms.

    Now all the villages located at different locations in Wai. Satara, Koregaon and Jaoli talukas of Satara District present agricultural crops are wheat, jowar, sugarcane. Sugarcane is cultivated in the canal benefited parts of the rehabilitated village. Most of the villages have got rocky and infertile and barren land.

15. Total land acquired : Land under submergence: for Dhom Lake: 5704 acres. (For Project: 2835 ha.)
16. Villages affected partly by the project : some of the families of the villages are still settled to the upward side of the dam. They are reluctant to leave their original locations. Percentage of such families is 15.
17. There are only 30 per cent of the villages rehabilitated at least with some facilities. The remaining are still fighting against many different social economic and agricultural problems.
18. Wai Taluka have changed its agricultural landuse along the sides of right and left bank canals. Canal irrigated part is fully occupied by sugarcane and vegetable cultivation. Only one-third part of the tahsil is benefited by canal irrigation.

    Underdeveloped and backward part of Wai taluka : The westward part is occupied by the Government for the project : Underground water level of central part of the taluka is increased by 3 to 5 feet.

    From Panchagani to Wai, the occupational structure is changing as the importance and fertility of agricultural land is increasing towards east. There were 13 villages to whom the market centre was Panchagani—a hill station. The villagers were migrated towards Panchagani in seasons to work as gardeners, coolies and many other occupations of the area.

    Rainfall of the area is decreasing eastwards.
19. The water level for the present of Dhom Dam Lake :
    731.66 Meters, 7th May, 1983.
    748.30 Meters, August, 1982.
    745.70 Meters, August, 1981.
20. Fully affected villages shifted elsewhere are:
    Asre, Asgaon, Velang, Vyahali, Chandavadi, Boriv, Govedigar, Mugaon, Bhivadi, Khavali, Vasole, Kondavali, Nhalewadi.
    Partly affected: Malatpur, Renavale, Dhom, Kondavale, Vashivali, Dhavali, Borgaon Bk. and Kh., Dahyar, Vaigaon, Golewadi, Partawadi, Panas, Vadoli.
21. Still there is S.T. Bus facilities to the upward part along the dam (lake) connecting the villages still remained in the original parts with proper upward locations; those are :
    Khavali, Dhavali, Mugaon.
22. Pachagani has got a very good site to observe the area of Dhom Lake : A man made beauty in the upper Krishna Valley.
23. It is observed that no complete village is resettled/rehabilitated as one unit or one village in any new location given by the government.

24. In the original location forest products were also important economic sources to them.
25. It is observed as 80 per cent of affected families are unhappy and not benefited at least what they have lost under the dam project.
26. Wai and nearby villages closed to canals are affected by increase of mosquitoes which give birth for disease like malaria.
2ˉ. Village Vyahali is rehabilitated at Kadegaon (Tal. Wai) at the westward side of Asale-Bhavani Temple Hill.
28. Village Chandavadi is shifted at Jamb Mal area in Wai Taluka near Bhuinj Sugar Factory, in the village area of Khadki, Tal. Wai. The village market centres are Bhuinj and Pachawad.
29. Old sites were well known for rice cultivation as they were to the westward, i.e., Konkan—part of Wai Taluka.
30. Area at the eastward of the Tahsil, is under cultivation of jowar, wheat, sugarcane, groundnuts.
31. Old location of Vyahali was to the south of Dhom along the right bank of Krishna, where at present the southward dam wall is constructed. (dam wall extension is south to north). Original location of Chandavadi was at the right bank of tributary Walki nadi, which meets to Krishna at the upward part of village Dhom Dam.
32. Vyahali is located at southern part of Kadegaon village to westward of Asale-Bhavani temple hill—an infertile land of northward slope of Vairatgad. It comes under group grampanchayat of Kadegaon.
33. Village Chandavadi is located on Jamb Mal under the village Khadaki of Wai Taluka and presently included in the group grampanchayat of Khadki, Taluka Wai.

APPENDIX III

Irrigation Projects of Satara District

THE district is situated partly in Bhima and partly in Krishna River Basin. The tahsils Khandala, Phaltan and Man are situated in the Bhima River basin and the rest of the tahsils are situated in the Upper Krishna Basin. The district has an area of 10492.0 sq. kms. (17° 5′ and 18° 11′ N, and 73° 33′ and 74° 54′ E). The River Krishna is the principal River. The normal rainfall of the district is 102.5 cm.

In Satara district, there are 1200 villages in which 889 are Kharif and remaining Rabi-crop-villages, but still this is not noted in the government record.[1]

Compared to the last 10 years, the year 1982 has a very little rainfall for agricultural farms. It is necessary to have scientific management of water through irrigation projects in view of getting minimum food crops.

Here in Satara district, also development of irrigation projects resulted in problem of rehabilitation of many villages.

At present, there are Urmodi, Tarale and Andhali Project in Satara District are important projects, which will be completed in the near future.

REFERENCE

1. Marathi Newspaper—Daily Sakal, dated 18th September, 1982, Saturday.

N.B.—Please see on next page for the table of "Irrigation Projects in Satara District.

APPENDIX III—*Contd.*

Irrigation Projects in Satara District

Name and location of the project	Dhom-Krishna project Wai Taluka	Kanher project Satara Taluka	Urmodi project Satara Taluka	Veer Dam village: Bhade (Tal. Khandala) (Veer, Dist. Pune)	Yeralawadi project	Ranand tank	Nher tank	Mhaswad tank	Tarali tank
1	2	3	4	5	6	7	8	9	10
1. Year of completion of the dam	1977	1984	1964	1966	1977	1958	1889	1901	1964
2. Estimated cost (in Rs. lakh)	3,421.74	2,910.54	17.16	584	273	30.18	8.25	21.28	13.95
3. Dam height in metres	49.98	50.34	5.17	36.00	19.56	19.32	22.56	24.33	4.88
4. Length of canals in km.	190	78.80	24.00	170.00	46.00	27.34	52.50	40.00	22.20
5. Maximum storage capacity in million cubic metres	382.32	286.00	7.5	278.33	56.80	7.11	9.12	3.80	28.54
6. Gross command area in hectares	42,392 (Satara)	12,745	1,822	22,65,501 (44,704 for Satara Dist.)	5,060	81.56	45.48	17,922	2,449
7. Culturable command area in hectares	33,913	10,195	1,700	1,81,195 (35,766 for Satara Dist.)	4 037	3,887	4,324	1,098	1,862

8. Irrigable area in hectares	28,829	9,182	1,417	26,400 (17,646 for Satara Dist.)	—	1,093	2,064	4,049	—
9. Area irrigated at present in hectares	1,298	—	123.00	4,056 (for Satara Dist.)	—	16.0	25	12	8
10. Rehabilitated villages	29	28	—	23,359 (15 013 for Satara Dist.)	—	76.0	1,564	2,376	151
Partly	15	18	—	—	—	—	—	—	—
Fully	14	10	—	—	—	—	—	—	—

Note : Koyana is a hydroelectric project, not included in the table.

APPENDIX IV

Data Collected from Rehabilitated Vyahali and Chandavadi, 1982-83

VYAHALI

Category landholdings (in acres)		No. of studied families	Population			Dependents					
			Male	Female	Total	Above 16 years of age group			Below 16 years of age group		
						Male	Female	Total	Male	Female	Total
A	1-3 0.40-1.21 hectares	22	82	78	160	9	39	48	28	33	61
B	3-5 1.21-2.02 hectares	8	51	42	93	4	25	29	24	17	41
C	5-10 2.02-4.04 hectares	4	16	11	27	1	10	11	8	1	9
D	10-15 4.04-6.06 hectares	1	5	3	8	–	2	2	3	1	4
	Total	35	154	134	288	14	76	80	63	52	115

Category		Total studied families	Earning population			Employed at					
			Male	Female	Total	Urban centres			In agricultural fields		
						Male	Female	Total	Male	Female	Total
A		22	45	5	50	20	—	20	21	—	21
B		8	23	—	23	16	—	16	12	—	12
C		4	7	—	7	2	—	2	5	—	5
D		1	2	—	2	—	—	—	2	—	2
	Total	35	77	5	82	38	—	38	40	—	40

Category	Families studied	Population in agricultural labourers			Land obtained (in acres) in rehabilitation			Total land-holdings at old villages (in acres)	Land for the present available above water level of Dam
		Male	Female	Total	Irrigated	Unirrigated	Total		
A	22	5	5	10	12.12	22.17	34.29	72.2	19.20
B	8	—	—	—	2.20	28.15	30.35	35	5.20
C	4	—	—	—	7.0	10.0	17.0	43	3.00
D	1	—	—	—	1.0	14.0	15.0	18	0.00
Total	35	5	5	10	22.32	74.32	97.64	168.2	28.00

APPENDIX *IV—Contd.*

Data Collected from Rehabilitated Vyahali and Chandavadi, 1982-83

CHANDAVADI

Category of Landholdings in acres	No. of studied families	Population			Dependents					
					Above 16 years of age group			Below 16 years of age group		
		Male	Female	Total	Male	Female	Total	Male	Female	Total
A 1-3 0.40-1.21 hectares	23	64	62	126	6	25	31	21	30	51
B 3-5 1.21-2.02 hectares	10	43	39	82	5	19	24	16	16	32
C 5-10 2.02-4.04 hectares	5	25	19	44	2	10	12	11	8	19
D 10-15 4.04-6.06 hectares	—	—	—	—	—	—	—	—	—	—
Total	38	132	120	252	13	54	67	48	54	102

Category	Total studied families	Earning population			Employed at					
					Urban centres			In agricultural fields		
		Male	Female	Total	Male	Female	Total	Male	Female	Total
A	23	37	7	44	17	—	17	18	4	22
B	10	21	5	26	8	—	8	10	—	10
C	5	12	1	13	5	—	5	6	—	6
D	—	—	—	—	—	—	—	—	—	—
Total	38	70	13	83	30	—	30	34	4	38

Category	Families studied	Population in agricultural labourers			Land obtained (in acres) in rehabilitation			Total land-holdings at old villages (in acres)	Land for the present available above water level of dam
		Male	Female	Total	Irrigated	Unirrigated	Total		
A	23	8	3	11	15.15	37.7	48.22	122.36	22.22
B	10	2	4	6	22.31	17.5	39.36	79.3	9.6
C	5	—	1	1	29.0	5.10	34.10	44.0	0.0
D	—	—	—	—	—	—	—	—	—
Total	38	10	8	18	67.5	55.22	112.28	245.39	31.28

APPENDIX *IV—Contd.*

1982-83

Name of the village	No. of studied families	Category landholdings in acres	DAIRY ANIMALS						Daily total milk productions in litres (present)	Dhom canal irrigated land in acres	Land obtained in time				Have you obtained sufficient agricultural land	
			At old locations			At new locations					Agricultural land		Settlement plots			
			Cows	Buffaloes	Total	Cows	Buffaloes	Total			Yes	No	Yes	No	Yes	No
Vyahali	22	A : 1 to 3	4	11	15	2	12	14	9	34.26	13	9	21	1	2	20
	8	B : 3 to 5	2	12	14	2	10	12	13	19.19	5	3	7	1	4	4
	4	C : 5 to 10	2	4	6	2	3	5	3	17.00	3	1	4	0	0	4
	1	D : 10 to 15	0	4	4	0	1	1	1	15.0	1	0	1	0	1	0
Total	35	—	8	31	39	6	26	32	26	86.5	22	13	33	2	7	28
Chandavadi	23	1 to 3	7	31	38	5	15	20	25	37.14	13	8	22	1	6	17
	10	3 to 5	8	24	32	4	11	15	28	29.6	7	3	10	0	3	7
	5	5 to 10	8	10	18	5	4	9	23	29.10	3	2	4	1	3	2
	—	10 to 15	—	—	—	—	—	—	—	—	—	—	—	—	—	—
Total	38	—	23	65	88	14	30	44	76	95.30	23	13	36	2	12	26

Note : For Chandavadi there were 322 sheep and goats, now remained only 5.

Village	No. of studied families	Category in acres	Your family budget maintainable or not?		Have you settled within your agricultural land		Do you feel rehabilitation profitable		Have you obtained agricultural land as per your choice?		Do you work in your own agricultural farms?		No. of families taking interest in dairy cattles
			Yes	No	Yes	No	Yes	No	Yes	No	Yes	No	
Vyahali	22	1 to 3	4	13	5	17	7	13	7	15	17	5	7
	8	3 to 5	1	7	—	8	1	7	—	8	6	2	6
	4	5 to 10	1	3	—	4	—	4	1	3	4	—	2
	1	10 to 15	—	1	—	1	1	—	—	1	1	—	1
Total	35		6	29	5	30	9	24	8	27	28	7	16
Chandavadi	23	1 to 3	8	15	3	20	—	23	3	20	2	21	16
	10	3 to 5	3	7	2	8	—	10	4	6	9	1	9
	5	5 to 10	1	4	1	4	2	3	2	3	4	1	3
	—	10 to 15	—	—	—	—	—	—	—	—	—	—	—
Total	38		12	26	6	32	2	36	9	29	15	23	28

APPENDIX IV—*Contd.*

Preferential vote for the help and for giving importance on the following items

	No. of studied families	Category in acres	For construction of temple	Medical centre	Help to poor sicks	Agree to invest money in			
						S*	G*	J*	B*
Vyahali	22	1 to 3	7	10	5	17	5	6	2
	8	3 to 5	3	2	3	7	2	2	—
	4	5 to 10	2	2	—	4	2	1	—
	1	10 to 15	2	1	—	1	—	1	—
Total	35		14	15	8	29	9	10	2
Chandavadi	23	1 to 3	7	10	6	23	11	10	8
	10	3 to 5	2	3	5	9	6	7	5
	5	5 to 10	2	2	1	5	2	5	3
	—	10 to 15	—	—	—	—	—	—	—
Total	38		11	15	12	37	19	22	16

*S : For Satyanarayana Puja.

*G: Dinner Party to all the villagers at the time of marriage ceremony.

*J : Investment for annual fair party.

*B : To kill animals for the sake of God.

Master Sheet of Dhom Dam Affected Villages (1971 Census)

Village	Total population	Total area (in acres)	Area under forest	Irrigated land		Unirrigated land	Culturable waste	Not available for cultivation	Occupied residential houses
				Well	River				
1	2	3	4	5	6	7	8	9	10
Fully affected villages (1971 Census)									
1. Asare	1,412	494	153	26	—	442	635	46	108
2. Asgaon	123	1,261	323	187	200	257	213	81	99
3. Velang	1,357	1,669	196	33	42	708	571	119	232
4. Vyahali	937	887	244	—	86	425	76	56	134
5. Chandavadi	314	201	—	31	30	89	27	24	42
6. Boriv	567	466	—	—	85	222	110	49	98
7. Govedigar	582	666	—	22	107	185	255	97	96
8. Mugaon	322	527	—	—	127	337	18	45	26
9. Bhivadi	608	651	171	40	83	158	170	29	89
10. Khavali	634	969	154	—	130	298	345	42	90
11. Vasole	1,075	2,453	1,177	44	169	467	542	54	197
12. Kondavali Kh.	284	581	140	—	105	82	220	34	48
13. Kondavali Bk.	542	913	364	10	56	323	138	22	103
14. Nhalewadi	377	548	—	—	51	230	184	83	67
Total affected	10,245	12,286	2,922	393	1,271	4,223	3,504	781	1,429

APPENDIX IV—*Contd.*

1	2	3	4	5	6	7	8	9	10
Partly affected villages									
1. Partavadi	459	756	285	—	30	159	252	30	58
2. Nandgaon	322	661	304	—	60	83	194	18	44
3. Panas	202	127	—	—	55	52	3	17	33
4. Vadoli	283	601	134	—	54	108	273	32	46
5. Malatpur	441	712	142	—	144	313	72	41	57
6. Renawale	889	947	133	26	159	250	332	47	122
7. Dhom	1,367	1,263	—	137	141	534	304	147	262
8. Kondavale	499	2,643	1,517	—	37	172	885	32	86
9. Vashivali	540	882	264	—	114	193	267	44	89
10. Dhavali	175	271	119	70	—	65	2	15	23
11. Borgaon Kh.	460	565	—	45	—	225	249	46	62
12. Borgaon Bk.	566	745	210	203	—	190	107	35	66
13. Dahyat	413	614	105	72	—	165	267	9	51
14. Vaygaon	579	736	165	6	58	212	245	50	70
15. Golewadi	344	1,135	410	—	37	95	527	66	43
Total	7,539	12,658	3,490	559	889	2,816	3,975	629	1,112

APPENDIX IV—*Contd.*

Master Sheet of Dhom Dam Affected Villages (1971 Census)

Village	No. of Households	Total Workers	Cultivators	Educated persons	Non-workers
1	2	3	4	5	6
Fully affected villages					
1. Asare	108	541	518	250	871
2. Asgaon	99	225	142	377	1,006
3. Veleng	237	567	542	338	790
4. Vyahali	135	273	84	134	614
5. Chandavadi	51	148	117	90	166
6. Boriv	121	238	196	174	329
7. Govedigar	108	261	212	131	324
8. Mugaon	54	76	61	70	246
9. Bhivadi	101	293	289	97	315
10. Khavadi	96	201	186	75	433
11. Vasole	207	313	293	186	762
12. Kondavali Kh.	48	129	118	72	155
13. Kondavali Bk.	103	262	240	108	280
14. Nhalewadi	75	168	142	120	209
Total	1,543	3,695	3,050	2,222	6,500

APPENDIX IV—*Contd.*

1	2	3	4	5	6
Partly affected villages					
1. Partavadi	86	115	147	108	294
2. Nandagane	56	179	142	62	143
3. Panas	39	105	103	26	97
4. Vadoli	46	82	51	83	201
5. Malatpur	76	108	78	111	333
6. Renawale	136	404	400	85	485
7. Dhom	262	422	229	560	945
8. Kondavale	97	100	97	32	399
9. Vashivali	99	139	89	56	400
10. Dhavali	31	46	35	36	129
11. Borgaon Kh.	115	121	120	46	339
12. Borgaon Bk.	97	157	154	96	409
13. Dahyat	82	198	171	45	215
14. Vaygaon	75	265	258	99	314
15. Golewadi	53	82	81	42	269
Total	1,350	2,523	2,147	1,487	4,965

APPENDIX V

List of Questionnaires (1 to 6)

1 & 2. Survey of displaced persons by Krishna Dhom project. (Geographical and socio-economic survey of Vyahali and Chandavadi).

1. (*A*) Family Background,
   (*B*) Position of agriculture,
   (*C*) Occupations,
   (*D*) Employments,
   (*E*) Labour-work,
   (*F*) Residential house: (village settlement),
2. (*G*) Family background,
   (*H*) Rehabilitation.
3. Interviews with the members of the administrative staff, workers in the Rehabilitation, Land-Acquisition and irrigation departments for Dhom Project.
4. Dhom Rehabilitation: Interviews with social and political leaders.
5. Interviews with the villagers of command area of Dhom Dam.
6. Interviews with shop-keepers and merchants of Wai/Bhuinj/Pachawad: effect of rehabilitation over the market centres.

QUESTIONNAIRES (1 to 6)

Indian Council of Social Science Research, New Delhi

QUESTIONNAIRE No. 1
1982-83

Survey of displaced persons by Krishna Dhom Project. (Geographical and Socio-Economic Survey of Vyahali and Chandavadi). (All the information provided by you will be kept strictly confidential).

(*A*) *Family Background*

1. Name of the village: (Rehabilitated)..
2. Occupation of the family...
3. Caste ....
4. Annual income in Rs..
5. Joint/Divided family...
6. ....

Sr. No.	Name of the family member	Relation with the Head of the family	Age	Education	Occupation	Monthly income in Rs.
(a)	Male					
(b)	Female					
(c)	Male & female below 16 years of age group					

7. Total Number of persons in the family: Male+Female=Total
8. Dependents below 16 years of age group: Male+Female=Total
9. Dependents above 16 years of age group: Male+Female=Total

(*B*) *Position of Agriculture*

Agricultural and obtained in Rehabilitation from the villages	*Distance from agricultural land with new village settlement in Kms.*	*Irrigated land*	*Non-irrigated land*	*Total land in acres*	*The name of original landholder*

1. Are you settled in your farm house? Yes/No.
2. Have you obtained agricultural land as per your choice in rehabilitation? Yes/No.
3. Have you been benefited by Dhom Canal irrigation? Yes/No. Area under canal irrigation: in hectares:..
4. Are the benefited villagers of command area giving you a good co-operation in the agricultural implementation work? Yes/No.

   (*i*) Explain the type of help you have obtained from them?.........
   (*ii*) If not, then why?...

5.	*Old village*		*New village*	
6.	*Land in ha.*	*Survey/Gat No.*	*Land in ha.*	*Survey/Gat No.*
	Irrigated			
	Unirrigated			
	Barren			
	Grass land			
	Total			
7.	*Trees you possess*	*Total No.*	*Trees you possess*	*Total No.*
	Chinch			
	Mango			
	Other			

8.	*Old village*.....................		*New village*..........................	
	Crops: *Cultivated area*	*Per ha. income in Rs.*	*Crops*: *Cultivated area*	*Per ha. income in Rs.*
(*i*)	Sugarcane			
(*ii*)	Turmeric			
(*iii*)	Vegetables			
(*iv*)	Wheat			
(*v*)	Rice			
(*vi*)	Jowar			
(*vii*)	Maize			
(*viii*)	Pulses			
(*ix*)	Oilseeds			
(*x*)	Other			
	Total			

9.	*Irrigation facilities*	*Well/ Canal/ River*	*Irrigation facilities*	*Well/ Canal/ River*

10. Wage Rates: For Female: Rs...............For Male: Rs..................
11. Suggest Development plans which are necessary to your new agricultural land:........................ Your problems and difficulties in agriculture:..

12. Have you obtained possession of agricultural land and residential plot immediately at new village area?
Possession of agricultural land: Yes/No.
Possession of Residential Plot: Yes/No.
13. Difficulties occurred at the time of possession of the land: ....
14. Have you obtain sufficient agricultural land in rehabilitation? Yes/No.
15. Indicate the iocations, where you have asked the government for rehabilitation: ....
16. For the present are you selling your agricultural products in the market? Yes/No.
Which products do you sell and where?..................................
17. Which are the products, you sell from your original agricultural farms? ....
18. Is there any agency, helping to you and providing informations regarding your agricultural work? Yes/No.
19. Do you work in your own agricultural farms? Yes/No. *OR*
Do you implement agricultural work of your farm through labourers? Yes/No.
20. Is there land available above the water level of Dhom Dam at your old village area? Yes/No. Cropwise cultivated land in hectares:..........................Annual income from old agricultural area in Rs...
21. Is it true that the Revenue charges collected from you for the land which is lost under the dam? Yes/No.
22. Due to large distance between your new residential shelter and agricultural land, do you find it difficult to concentrate on your house property and your family matters? Yes/No. How do you take care of your agricultural farms, being a location away from your house?...
23. The places and names of the villages, where your family members, relatives bifurcated and resettled:

Name of the village	*Family members/ relatives*	*Land obtained in ha.*			*Survey/ Gat No. of the land*
		Irrigated	*Non-irrigated*	*Total*	

24. In rehabilitation have you obtained land from the village of your relatives? Yes/No.
Name of the relatives with relation...

(*C*) *Occupations*

(Animal rearing and other occupations).

From Old Village	*In the New Village*
1. Name of the occupation	
2. Annual income Rs.	
3. Dairy Cattles Total:	
4. Dairy milk production in Litres:	
5. Other animals:	
6. Total	7. Total
8. Hens: Total:	
9. Other occupations with their annual income:	

10. Daily requirement of milk to your family: in Litres:..................
    Annual expenditure for milk requirement in Rs..........................
11. Annual Expenditure on fertilizers and manures: Rs....................
    Fodder in Rs..
12. Cattle shed is connected with inside to your house/closed to your kitchen room.
13. Which business have you started from the amount of compensation you obtained in rehabilitation?...

(*D*) *Employments*

Employed persons of your family:

1. Total number of family members employed elsewhere:
   Male +Female =Total

Name of the person	*Address of his institution*	*Designation*	*Monthly total salary in Rs.*

2. Have anybody from your family benefited being dam affected person for getting facility of employment? Yes/No.

(*E*) *Labour-work*

1. Are you agricultural labourer? Yes/No.
2. No. of family members working as a labourer:
   Male +Female =Total

3. *Facilities of old village House/Hut/Land/Other*	*New village House/Hut/Land/Other*

(*F*) *Residential House* (*Village Settlement*)

1. Are you residing for the present in your own or rented house? ......
2. Was your faith/is your faith connected at your old/new residential home?
3. Have you obtained location of your house plot according to your choice? Yes/No.
4. Has the Govt. helped you in connection of providing transportation facility to carry your house materials? Yes/No.
5. Have you constructed your residential shelter on your own cost or with the help of loan taken from the Govt.? Yes/No.
   Total expenditure Rs..................Govt. help/loan in Rs..............
   Loan sanctioned from...............Still unpaid loan in Rs...............
6. Are the families residing very close to you or in a Wada type of house structure as they were, in the old location? Yes/No.
7. Location of Buddhist settlement: Within your new settlement of the village/outside of your new village.
   Distance (in metres).................... To the side of ....
8. Is there facility of common lavatory blocks in the new village? Yes/No.
9. Is there suitable cremation ground? Yes/No.
10. Is there separate "Gram Panchayat" and village Co-operative Society to your new village? Yes/No.
    Your new village comes under "Grampanchayat" of village.........
    and village Co-operative society of....................village.
11. Have you taken loan from your old village Co-operative society? Yes/No.
12. Have you paid the loan? Yes/No. Have you taken loan from the new village society? Yes/No. Rs...
13. Are the religious and traditional functions like Kirtans or Bhajans continued in the new village ? Yes/No.
14. How many of your relatives are residing at the old village? Total No. of families..................................Names of the Head of families ....

QUESTIONNAIRE No. 2

(*G*) *Family Background: Economic Position of Your Family*

1. Rehabilitation resulted to your family life, as it is improved/disturbed/and scattered.
2. Have you maintaining your family budgets and agricultural accounts? Yes/No.
3. Is there any "court matter" for which you have to fight against anybody? Yes/No. Give details of it:..

4. Have you celebrated marriage of your son or daughter at the new location? Yes/No. Have you taken dowery in the marriage of anybody in your family? Yes/No.
   Total marriage expenditure: Rs...
5. Being a affected person of Dhom Dam are your relatives ready to help and to keep for maintaining family relation with you? Yes/No.
6. Are you ready to allow your female family members to take education/to get employment service/to participate in the village social work? Yes/No.
7. Is there artistic skill of any woman in your family, in sewing drawing, painting texture and embroidery and in agricultural work? Yes/No.
8. Do you expect that your daughter should get her husband from rural/urban area?
9. Is it possible to maintain your family expenditure? Yes/No. Any reason with remedies to maintain your family budget.................
10. Is there anybody who is always found sick under any disease in your family? Yes/No. Indicate details of it:...............................
11. Is there anybody of your family taking interest for helping in social work of the village? Yes/No.
12. Give preferential vote in the following items as per your choice, or indicate your first preference to the following social work:
    (*a*) Help for construction of a temple, (*b*) Help for construction of a dispensary or hospital, (*c*) Help to the sick handicapped person for medical purposes.
13. Do you believe in caste spirit? Yes/No.
14. Do you agree to accept, a simple unexpensive method of celebrating marriage of your son, rather than expensive showy nature of expenditures? Yes/No.
15. Do you agree to investment of amounts in the following items:
    (*a*) Satyanarayan Puja. (*b*) Marriage party to your village. (*c*) Dinner party in annual fair. (*d*) Mothi Bai expenditure and sacrificing goats for religious purposes.
16. Is it necessary for you to purchase foodgrains for your family because your own produce is insufficient? Yes/No. Give details with its total annual expenditure:..
17. Have you planned to save the amount for future expenses (like education) for your children? Yes/No.
18. Total unemployed in your family:
    Male +Female =Total

19. Annual grocery expenditure in Rs.......................Expenditure on cloths in Rs.........................Expenditure in festivals and annual fair Rs..
20. Children Education of the family:

S.S.C.		H.S.C		Graduates		Total		
M	F	M	F	M	F	M	F	Total

M=Male; F=Female

21. Are the village Talathi and Gramsevak, helpful and co-operative with you? Yes/No.
22. Have you supported for corruption by giving to anybody some amount as a bribe? Yes/No. Give details of it:.........................
23. Have you obtained a good co-operation from your village sarpanch or Police Patil? Yes/No.

(*H*) *Rehabilitation*

1. Do you feel rehabilitation is more expensive, unprofitable or beneficial? Yes/No.
2. Do you like new location? Yes/No.
   Give reasons..
3. Year in which you have settled in the new location:.....................
   date:..........................Give the year in which you have obtained possession of your agricultural land.................date:..............
4. Year in which your agricultural land is benefited by canal irrigation:
   Year:...........................Date..
5. Is there anybody who have taken cognizance of your rehabilitation problems? Yes/No. Give details of it:......................................
6. Explain the facilities provided to you by the govt.: in rehabilitation.
   Economic facilities: loan remission:
7. Help that you have obtained at the time of shifting and settling at new location: (*a*) In the form of amount: in Rs..........................
   (*b*) In the form of materials:...............(*c*) Other help:..............
8. In the work of rehabilitation, have you been cheated or deceived by anybody? Yes/No. Give details of it..................................
9. In your new village location, have you taken interest in tree plantation? Yes/No. Name of the trees..................................

10. Which were the assurances of providing facilities given to your family at the time of resettlement by the govt.?........................
Whether the facilities are provided to you? Yes/No.
11. Was there "Morcha" for demand of rehabilitation problems arranged from anybody of your village? Yes/No. Have you participated in it? Yes/No. Are you benefited by it? Yes/No. Date and reason for "Morcha"..
12. Was the amount sanctioned to your family in rehabilitation misused or utilized properly by anybody of your family?..............
What type of loss occurred due to misuse of the amount?...........
Due to improper utilization of the amount, is there anybody from your family who has spoiled his life by bad habits? Yes/No.
13. Merits and demerits of rehabilitation:......................................
Agricultural difficulties:...........................Economic and social difficulties:.........................Family problems:.......................
14. Is there educational facility (Primary Education) provided to your new village? Yes/No. Which are the other facilities required to you and to your village settlement?..
15. Are you a member of any village or social fund collected for the betterment of the village? Yes/No. Title of the 'Fund'.................
Amount paid in Rs...
16. Your relative residing within or closed to their agricultural farms:

Relatives	*New village of the agricultural farms*	*Survey No.*

17. Are your relatives affected by irrigation project? Yes/No. Name of the relatives:.........................Village:...............................
18. How was the picture of your original village? Explain your feelings about it, into two sentences:..
19. Explain into two sentences about the environment of your new rehabilitated village:...
20. Give details for the amount obtained to your family in rehabilitation:
(*a*) Compensation for acquired agricultural land in Rs.................
(*b*) Compensation for acquired house in Rs ....
(*c*) Compensation for acquired trees in Rs................................
(*d*) Other if any...

21. Compensation obtained to you: in the following items:
    (*a*) Agricultural land in hectares..
    (*b*) Residential plots: in sq. metres:..
22. Excluding the government help, obtained to you, is there any help extended up to you from any institution or any village of command area? Yes/No. Explain in one sentence:...................
23. Have you obtained agricultural land and residential plot at one location? Yes/No.
24. Explain in two sentences, of about your expectations from the government, in the work of Rehabilitation by irrigation project: ....
25. Are there Political/Social/Racial/Other conflicts always occurred in your village?...
26. Other information provided by a person of dam affected family......
27. Surveyor's observations and special remarks about the family.........

Signature of the Surveyor/Investigator

QUESTIONNAIRE No. 3

Indian Council of Social Science Research, New Delhi

Interviews with the members of the administrative staff, working in the Rehabilitation, Land Acquizition and Irrigation departments for Dhom Project (with special reference to Vyahali and Chandavadi: selected villages).

(Please do not mention your name and designation in this questionnaire).

1. In which department, do you work?.......................................
2. What difficulties you have faced while executing and implementing decisions taken by the government in relation to rehabilitation work?
3. By providing false statements and wrong informations, some of the dam affected families have taken maximum benefits from the govt. Do you agree for this? Yes/No. Give an example.....................
4. At the time of acquiring land and settlements, from catchment area, do you observe that all the affected villagers were opposed to your action? Yes/No.
5. Do you agree that in rehabilitation, the families of shifted villages, have obtained sufficient land compared to that of original village? Yes/No. What are the difficulties you have faced by giving insufficient land to the dam affected families?..................................

6. Have you given proper consideration to the opinion of affected landholders at the time of giving them agricultural lands and residential plots? Yes/No. Have you considered at the time of distributing agricultural land to the affected families, by giving them new locations closed to their relatives? Yes/No.
7. Do you agree that the rehabilitation work is quite satisfactory of Dhom project? Yes/No. Give reasons if the work is observed to you as unsatisfactory with names of villages:...........................
8. Suggest your remedies and ways to solve the problems of dam affected villages...
9. Which is the difficult problem you feel at the time of rehabilitation work?...
10. Suggest your ways, while satisfying the complaints, raised before you against construction of irrigation project and improper rehabilitation:.. ....
11. Suggest a suitable way for maintaining good co-ordination in between Government agencies, political leaders and rehabilitators:......
12. What is the difference do you observe while obtaining co-operation from the shifted villages and the villages of command area who have given their agricultural land and settlement land to the affected villager's?...
13. Was you responsible for sometime to delay the work of Dhom Rehabilitation? Yes/No. What is your explanation for such delay of work?..
14. Suggest your suitable way for quick administrative and other field work:..
15. Have you been benefited by local political leaders at any time in solving the problem of rehabilitation? Yes/No.
16. Have you obtained a good co-operation from your office staff? Yes/No.
17. Do you follow and accept the suggestions from the experts, rather than the dam affected villagers?...
18. Do you feel it proper that the economic and other help, given to the dam affected persons is sufficient? Yes/No.
19. Do you agree that the agricultural land given to the displaced persons on a long distance from residential plots, is unprofitable? Yes/No.
20. Is it necessary and affordable to provide constructed buildings at new locations, rather than to give amount for constructing buildings? Yes/No.
21. Is it more easy of rehabilitation of a village rather than the shifting work of a village?...

22. Is it also easy to resettle a complete village at a location? Yes/No. Mention difficulties you have observed:..................................
23. Were there such demands from the villages for completion of Dhom Project? Yes/No. Villages supporting for the project..........
24. Give reasons, from the villages, which were against Dhom Project with their names:..
25. Explain the unsolved difficulties of Dhom Project Rehabilitation:..
26. Do you observe that the landless poor families are properly settled at new locations?...
27. Give informations regarding to those needy dam affected families are still not resettled and rehabilitated by the government : villages and total number of families:

    Chandavadi: Vyahali: Other villages:

28. As per your knowledge, indicate the help given by the government to the rehabilitated villages:...

Village	*Help in the form of materials*	*In the form of amount*	*Other help*

29. Do you observe that after the acquisition of land for the project, the landholders have still to bear the land revenue charges? Yes/No. Give examples:...
30. Do you feel the compensation given to the dam affected families, is suitable and proper, considering the present actual price of the land? Yes/No.
31. Were there any obstructions observed to you regarding land acquisition from villagers of command area? Yes/No. Give details of it:..
32. Is it necessary as per Dhom and Kanher Dam, "Pathakhal Borkhal" Project should be constructed being an important need of irrigation? Yes/No.
33. Do you feel that the rehabilitation work is a distressed, discriminative, and difficult work for your consciousness? Yes/No.
34. Give name of the village of "Dhom Project," rehabilitated satisfactorily:...
35. Is it sufficient that government has provided proper economic and other help to the dam affected families? Yes/No.
    Do you agree that the help is more than sufficient? Yes/No.

36. For taking benefit of proper rehabilitation, do you observe that such affected person came to your contact by using some political or other power? Yes/No.
37. Because of your help, tired affected families are resettled properly and become stable. Have you observed such families at the time of rehabilitation? Yes/No. Give details with their names and address:..
38. At the time of shifting the villages, do you observe their crying with sorrow? Yes/No.
39. To give land for the project is a great forsake. Do you observe that this feels very grateful to the villagers of command area? Yes/No.
40. Do you feel that project affected families have taken unlimited and more benefit from the Government? Yes/No.
41. Do you agree to the opinion that rehabilitated families have obtained rocky, infertile and insufficient land? Yes/No.
42. Is it true that due to personal relations and political impressions, some of the dam affected persons have taken benefit of obtaining agricultural land closed to their new settlement location? Yes/No.
43. It is a tendency of the villagers, always to show very little annual income from agriculture and from services of their family members. Explain your opinion:..,......
44. Indicate the names of five villages who are benefited by Dhom Canal Irrigation which have never lost their agriculture land for rehabilitation?...
45. The complaints and other applications related to rehabilitation are not immediately accepted by the government officers. Give your opinion on it:..
46. Is it not essential to provide certificate to dam affected person, after his rehabilitation at the new village? Yes/No.
47. How to cultivate scientific approach towards agricultural development in the minds of rural people?...
48. Do you feel quite improper and inhuman at the time of shifting and removing dam affected families, from old to new sites of their villages? Yes/No.
49. If you had been in the unsheltered place of dam affected person, explain in one sentence, what you would have done?.................
....
50. Suggest the remedies for the development of affected persons:.........
51. Indicate your views, plans in relation to village development:.........

52. After shifting up to the new settlement, the rehabilitated farmers take 2 to 3 years for settling. Do you feel that this period is unproductive and wasted period ? Yes/No. Suggest a remedy of it:..

QUESTIONNAIRE No. 4

Dhom Rehabilitation: 1982-83. Interviews with Social and Political Leaders

1. Have you worked under any rehabilitation committee of Dhom Dam? Yes/No.
2. Do you agree that the rehabilitation work of Dhom Project is satisfactory? Yes/No. If not, why?...
3. Do you feel the work is useful of any committee, appointed for rehabilitation? Yes/No. Why?...
4. Suggest your remedies to solve the problems of rehabilitated villages?...
5. The delay of the work lies with the hands of the members of the staff of the government. Is it true? Yes/No. Your suggestion to remove such drawback of delay in the work:..........................
6. Is there any complaint put forth before you by dam affected persons? Yes/No. Give in details about the complaints:................
7. Your remedies to solve the complaints:..................Give a brief account of your personal work you did in relation to rehabilitation:..
8. Have you joined in a morcha of dam affected persons, demanding their requirements? Yes/No. Give details of it:.......................... Date of Morcha.....................Demands of Morcha.................
9. Have you observed a good co-ordination in between local political leaders and the staff-members of the rehabilitation office? Yes/No.
10. Do you feel that the help and compensation given by the government to the Dhom dam rehabilitators is sufficient and proper? Yes/No.
11. Is it unprofitable to the rehabilitators that there is a great distance in between agricultural land and residential plot? Yes/No. Suggest your remedies on it:..
12. Is it true that sufficient land is given in compensation compared to original land to the dam affected families? Yes/No.
13. Is resettlement very easy than the work of shifting of village? Yes/No.
14. Mention the unsolved problems of Dhom rehabilitation? Yes/No.

15. Do you feel the necessity of construction of Pathakhal Borkhal Project? Yes/No. (Now this project is cancelled by the Govt.).
16. Nobody took a note of complaints and demands of project affected persons. Is it true? Yes/No.
17. Do you follow and accept the views and plans on any matter, suggested by experts and scientists? Yes/No.
18. Do you agree that the rehabilitators have obtained infertile, barren, and insufficient land? Yes/No.
19. Is it true that due to political impact and relations with political persons some families have taken benefit of obtaining agricultural land near the new village? Yes/No.
20. Police and other government officers do not take proper congnizance about the complaints put forth by the dam affected persons. Explain your views on this:..
21. Is it necessary to work for dam affected families by collecting economic and other help from the benefited villages of command area? Yes/No.
22. Would you like to contribute to the fund for the rehabilitation work? Yes/No.
23. Do you agree, from the villages, the unnecessary expenditures on annual fair, Ganesh festivals, should be minimised in such a way so that the remaining amount can be given for rehabilitation work? Yes/No. How this opinion will be extended up to the villagers?...
24. What will be your role, if you would have been in the place of rehabilitators? Explain in a sentence:.,....................................
25. Suggest a successful way for stability of healthy and economically sound co-operation in rural areas?..
26. Explain a way for proper utilization of energy from unemployed educated person:...
27. Do you agree to give equal importance to the following two items?
    (*i*) Compulsion of shifting the village for any project, and
    (*ii*) Compulsion on family planning and plant cultivation: Yes/No.
28. Your opinions and suggestions about rehabilitation of Dam affected villages: if any..
    Interviewer's Name and Address:..

QUESTIONNAIRE No. 5

Interviews with the villagers of Command Area of Dhom Dam

Name of the landholder:........................Age:...........................

Number of family members: Male +Female =Total

1. Village:...
2. Land: irrigated +non-irrigated =Total (in acres)
3. Have you benefited by Krishna Dhom Canal Irrigation? Yes/No. Your total irrigated land in acres:..
4. How much land is acquired from you for rehabilitation? Acres:..... Survey Nos...
5. Amount received as compensation from the government: Rs.........
6. Name of the person to whom your agricultural land is given.........
7. Have you lost your land under canal construction? Yes/No. Land in acres:..
8. Compensation obtained for the land acquired under canal route: in Rs...
9. Due to Krishna Dhom canal irrigation are you benefited of gaining profit from agricultural production? Yes/No.
10. Crops you cultivated:...

	Before canal Irrigation	*After canal irrigation*
Crops		
Cultivated area in acres		

11. Your annual income in Rs.
    Land Revenue
    Charges in Rs.
12. Is Canal Supervisor giving you proper co-operation to irrigate your land? Yes/No. If not, why?..
13. Mention profitable crops for your irrigated land which yields maximum income to you: Sugarcane/Turmeric/Vegetables/Other crops.
14. Are the absorbed rehabilitators of your village mixing with your family environment? Yes/No.
15. Is there co-operation in between you and the rehabilitated families for agricultural work? Yes/No.
16. Have you obtained proper compensation of your land? Yes/No.

17. What is the reason that you were not agree to give your land to the rehabilitators?..
18. Have you received a good co-operation from the government officers, at the time of handing over your land to the L.A. department? Yes/No.
19. Are you deceived by somebody in the rehabilitation work? Yes/No. If yes, then indicate his name and designation:..........................
20. For what purpose you have invested compensation amount obtained to you? Household expenditure/purchase of land/marriage function/agricultural development work/medical charge of sick persons in your family/other reason:..
21. Why particularly your land is selected for acquisition?..................
22. Though some part of your land is acquired by the government, is the remaining land turned fruitful and profitable to you due to canal irrigation? Yes/No.
23. Is it grateful, that due to shifting of dam affected villages, you are benefited? Yes/No.
24. Do you agree to give economic help, if contributed and collected for resettlement of dam affected villagers? Yes/No. Explain your plan of help to them.
25. Is it necessary that the villages of command area should also come for help to the rehabilitators? Yes/No.
26. If you would have been in the place of affected person. Explain in one sentence that what would be your role in this connection:........

QUESTIONNAIRE No. 6

Interviews with shopkeepers and merchants of Wai/Bhuinj/ Pachawad about the effect of rehabilitation over the market centres

1. Name of the shopkeeper/Merchant:..
2. Type of shops: Kirana/Stationery/Hotel/Cotton Hosiery/Agro-productions/Fertilizers/Goldsmith/Sewing/Other.........................
3. Indicate, name of the rehabilitated villages from which the consumers are visiting to your shop:................./dam affected villages were coming to visit your shop..
4. Rehabilitation of Dhom project is profitable/unprofitable to your business in the market. Yes/No.
5. In which season do you have more scope and profit to sell your goods? Annual fair/Festival/Marriage season/Other season:...........
6. Please indicate your shopping position that the consumers of your shopping centre are at present increased or decreased? Indicate the

ways you have followed for the development of your shopping centre ....

7. Which agricultural productions are entered in your market centre from the resettled/dam affected villages?..................................
8. Do you agree that 'at increasing rate of population, no decrease in consumer's strength is observed to you? Yes/No.
9. Your income, compared with past two years, is at present increased. Yes/No. Give reason for it:..
10. If an amount collected for the help for dam affected villagers, will you be able to share in it? Yes/No.

Index